Dubuque during the California Gold Rush

Dubuque during the California Gold Rush

When the Midwest Went West

Robert F. Klein

The History Press

Published by The History Press
Charleston, SC 29403
www.historypress.net

First published 2011

Manufactured in the United States

ISBN 978.1.60949.491.9

Library of Congress Cataloging-in-Publication Data
Klein, Robert F.
Dubuque during the California gold rush : when the midwest went west / Robert F. Klein.
p. cm.
Includes bibliographical references and index.
ISBN 978-1-60949-491-9
1. Dubuque (Iowa)--History--19th century. 2. Gold miners--Iowa--Dubuque--History--19th century. 3. California--Gold discoveries. 4. Dubuque (Iowa)--Biography. 5. Migration, Internal--California--History--19th century. I. Title.
F629.D8K54 2011
977.7'3902--dc23
2011044108

Other than the gold, I don't like the country near as well as Iowa.
—James Langton, November 26, 1849
Miners' Express, *February 6, 1850*

Contents

Preface

This book arose from my discovery of some seventy letters written by forty-niners who left from Dubuque, Iowa, to join the mania popularly known as the California gold rush. I became curious to know how the Dubuque community of 4,000, with an estimated 1,200 to 1,300 men involved in some facet of mining, coped with the loss of some 500 men to gold fever.[1]

Descriptions of mid-century Dubuque, the area's lead mining industry and a brief overview of the California gold rush provide documentation and context for this important westering experience. As a whole, a vivid picture appears of the nexus between miners from the nation's most important lead-mining area at that time and the California gold rush.

In the years from 1848 to 1852, sending and receiving letters was a dicey proposition. Formal mail stations and routes had not yet been established between far-off California and the United States. The Pony Express was not established until 1860, and getting correspondence back and forth from "home" to California required a great deal of patience, along with a good bit of ingenuity.

It is said that twenty-five thousand letters to the miners arrived in San Francisco in the summer of 1849. Not surprisingly, long lines of potential recipients formed to see if they would make a strike at the post office. Ever an enterprising group, individuals offered to stand in the long line for others—for a fee, of course. Others simply bought their way to a place near the head of the line, while still others sold coffee and pastries to those standing in the queue.

As expected, Dubuquers also encountered difficulties in corresponding back and forth from California to Iowa. William Bothwell was resigned to the poor service, noting the high cost of one dollar per letter, with very little probability of delivery. James Langton tried an informal yet practical approach. He identified a man going to New Orleans and entrusted him to take his letter there, from which place his letter might be more reliably mailed. And then there was Henry Gooddaire with this grumbling description of mail service:[2]

> *Letters directed to San Francisco frequently lay in the office, and they cannot be got out without some considerable trouble and loss of time, there being so many there that there will be hundreds waiting to take their turn to get to the office. The best way of sending letters is by private conveyance, if possible. It generally costs $1.00 a letter from here to the States, to be mailed there, and then they will go without delay; and $1.00 is not missed here in this gold country.*

Yet many of the letters from the Dubuque forty-niners did get through to family and friends. At least one of the writers, William H. Merritt, likely wrote with the intention of having his letters published, as he was the proprietor of the *Miners' Express*. For all the writers, it was a way to make contact, not only with family and relatives, but also with the many friends and acquaintances in the still relatively small community of Dubuque.

The local newspapers were pleased to publish the letters because of the intense interest in the gold rush by their readers. It was also seen as a service to readers, in that these letters often provided useful information to those contemplating joining the thousands already on the trail to California. It must be remembered that at this time, newspapers were the only mass medium available, and letter writing was the only form of personal communication, there being yet no telegraph from the West or other, more modern inventions.

Much of the research to locate the letters of the Dubuque forty-niners was conducted in the local newspapers, especially the *Miners' Express* and the *Dubuque Tribune*. Each of these has a record of multiple minor name changes. The names as given here will be used throughout.

Today, we can still hear the voices of the Dubuque forty-niners through their own letters and in their own words regarding the hardships encountered in their travels: the weather; care for their friends on the trail; concerns for their livestock; shock at prices for necessities; dismay at the long, arduous days; and sadness at the death of friends and fellow travelers from Dubuque.

What commenced as high adventure rather quickly turned into a rugged expedition trying the mental patience and physical strength of the most sturdy of these forty-niners. Many came to see their participation in the California gold rush as a hard and difficult time in their lives. Fortune did not favor everyone.

Gratitude for assistance in this work is extended to Mike Gibson, archivist, Center for Dubuque History at Loras College; Kristen Smith, information services librarian, Loras College; DiAnn Kilburg, InterLibrary Loan librarian, Loras College; Brooke Deely, PhD, director of women's studies, University of St. Thomas, Houston, Texas; Patrick Brunet, retired librarian/author; Mary Bennett, State Historical Society of Iowa; Susan Hellert, History Department, University of Wisconsin, Platteville; Heidi Pettitt, Loras College Library; Gerda Preston Hartman, local historian; Peter Harstad, PhD; Linda Mathewson; Terry Grant, photographer; Tom Klein; and Jim Swenson.

Introduction

This country is good for nothing—except its gold.
—James Fanning, undated letter, Miners' Express, *May 15, 1850*

In the years 1849 and 1850, the lead miners of the Dubuque, Iowa region in the Upper Mississippi Lead District and throughout the United States—and, indeed, the world—were seized by "gold fever." The Dubuque forty-niners who left house and home, comfortable living conditions and even positions of esteem in their community joined in this historic migration to California. Whatever their personal reasons may have been, or if just caught up in the frenzy, they joined the trek across the great Trans-Mississippi West to California with the prospect of ready and immense riches in the gold fields there.

Given the magnitude of the movement to California, it seems natural that it would have had an effect on Dubuque and the Upper Mississippi Lead District. At the time of the discovery of gold, Josiah Conzett later recalled in *Recollections of People and Events of Dubuque, Iowa*, "about half the town was engaged in mining for lead." This was the period of peak lead production in the region, and Dubuque was on its way to becoming the population center of the area.

The loss of lead miners leaving for the California gold fields would likely have had less of an impact on Dubuque, with its larger population, than in the smaller nearby mining villages in Wisconsin and Illinois. The reason for the disproportionate effect was that Dubuque, located on the west bank of the Mississippi River, was in a better position to offset its population loss with continuing immigration gains and a larger, more diverse economy. The smaller surrounding communities would have been more severely affected, as lead mining was often their sole economic engine.

This work presents an overview of the Upper Mississippi Lead District and Dubuque; the arrival of the news of the discovery of gold and the local reaction to that news; preparations to leave Dubuque and the Dubuque forty-niners' journeys to California; their initial euphoria and ultimate reconnection with reality; and finally, results brought to Dubuque and its citizens as a result of the gold rush.

Chapter 1

DUBUQUE

The Early Years

The presence of lead in the Dubuque area had long been known to the Native Americans living there, but for many years it was of minor importance to them. By the latter half of the 1700s, however, the exchange in lead increased as these people showed a greater desire for the trade articles provided by Euro-Americans. As a result, the mining of lead intensified, and something of a commercial industry developed.

Whites became aware of the abundance of the mineral at least as early as 1655 through Pierre Radisson and Medard Groseilliers. Pierre Hennepin's map of 1687 shows lead mines in the area, and Nicholas Perrot began mining operations in 1690.

Julien Dubuque, a Frenchman from Trois Rivières in Quebec, Canada, arrived in 1788 and fashioned an agreement with the Sauk and Fox tribes allowing him to work the lead mines in an area around what is now Dubuque, Iowa. No specific limits were stated in this accord, but Dubuque later claimed that his "lease" stretched from the Little Maquoketa River in the north to the Têtes des Morts Creek in the south, comprising a frontage on the Mississippi River of some twenty miles and to a depth of about ten miles back on the west bank. His mining efforts achieved mixed results. He shipped the success of his labors south on the Mississippi to Missouri, where he dealt with and contracted substantial debts to the St. Louis trader August Chouteau.[3]

Subsequent to Dubuque's death in 1810, the Native Americans living in the area regarded his passing as the termination of the "lease" they had made with him. Accordingly, they asserted sovereignty over their traditional lands

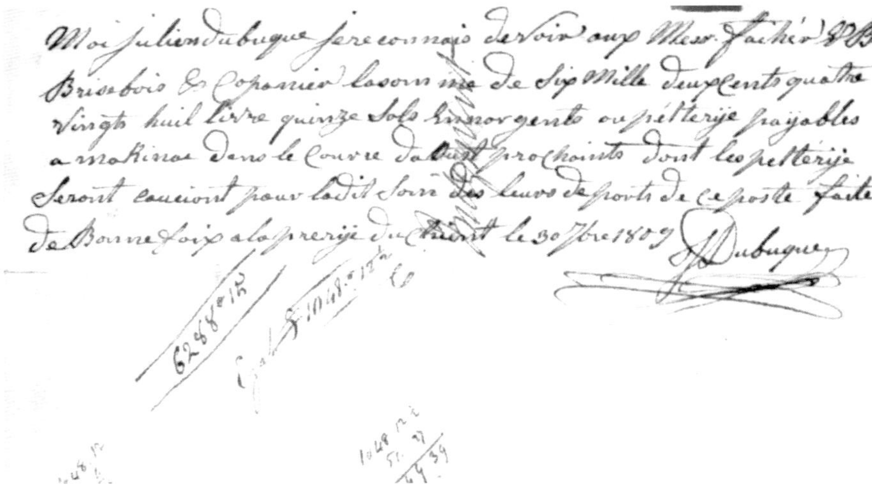

Julien Dubuque's contract of indebtedness, September 30, 1809. *Courtesy of the Loras College Library.*

and were successful for a number of years in keeping white entrepreneurs out of this region on the west bank of the Mississippi.

In the early 1830s, Caleb Atwater described the mines as the richest in the known world. A few years later, geologist David Dale Owen investigated this lead region in 1839 and proclaimed it as one of the richest mineral regions in the world. By 1852, the Upper Mississippi Lead District was producing some twenty-six million pounds of lead annually. This output was nine-tenths of the total produced in the United States and made up 10 percent of the world supply. It is clear that this was a thriving and important industry for the entire country.

During the 1820s, Galena, Illinois, had become the key outpost for the mining region, while the Iowa lands were still in possession of the Native Americans. Nevertheless, in the later 1820s and into the early 1830s, whites were engaged in unsanctioned exploration and lead mining on lands on the western side of the Mississippi.

The U.S. government, in order to preserve the tribal rights, sent troops to Dubuque's Mines, as they were called, from Fort Crawford at Prairie du Chien to roust out these "sooners" and return them to the eastern shore of the Mississippi. Even so, the choice locations for prospective lead mining on the Iowa side of the river had been identified.

The Black Hawk War arose in 1832 and was concluded on September 21 of the same year with a treaty in which the U.S. government gained control of six million acres of land in eastern Iowa in exchange for $640,000, approximately ten cents per acre.[4] White settlement was officially sanctioned as of June 1, 1833. From that time on, a much more concerted effort was made to mine the lands on the west side of the Mississippi.[5]

The earliest immigrants working the lead mines in the Upper Mississippi Lead District were a young, hardy, dynamic group of people who were invigorated by an adventurous spirit and the prospect of making money. At the beginning, they had relatively few creature comforts, some even wintering over in "badger holes." It could not have been an easy life.

A few short years later, what had been Dubuque's Mines officially became the town of Dubuque in 1837. By 1846, following the achievement of statehood for Iowa, it had become a small city on the move, as Josiah Conzett attests:

> [Dubuque in 1846] *was mostly all built of frame and log buildings with a few blocks of 2 story brick store buildings. These were all below 5th on Main St., with here and there a brick residence such as Ed Langworthy's on the corner of 14th & White, L. Langworthy's on the corner of 12th & Iowa, Judge T.S. Wilson's at 16th & White, the Blake house on Mineral St., the old Court House, the brick house on 7th & White and the one a little below the Jefferson House. Except for the last three, the others were then in the suburbs and nearly out of the town limit.*[6]

Two years later, the news that was to transform the nation struck Dubuque. Gold had been discovered in far-off California, which was not then part of the United States. Chandler Childs provides his assessment of the impact on the city. He identifies those who ventured into that comparatively undiscovered land on the Pacific coast as mostly young men, among whom were farmers, miners, clerks, merchants and some capitalists:

> *The effect of this emigration from Dubuque was not such as was calculated to encourage those who remained behind. After the departure of those who went west, business became flat, stale and unprofitable. No inconsiderable sums were expended by the adventurers in the purchase of outfits, tools, horses, etc., but with them once more departed the prosperity which had for several years previous coquetted with Dubuque. Mining, while not entirely abandoned, was engaged in at intervals and but carelessly prosecuted, and this interest did not revive to any appreciable extent until 1855. Emigration almost entirely ceased. The area of cultivation was measurably reduced,*

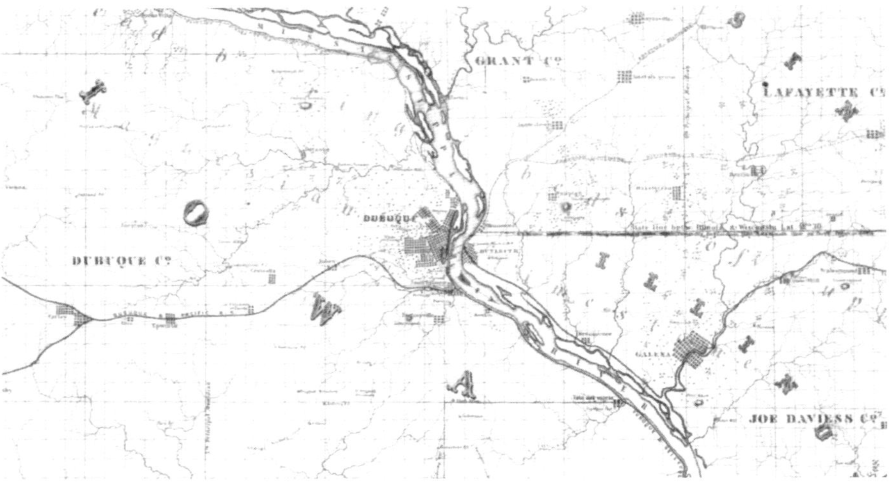

Mineral Region of Dubuque & Vicinity. 1858. De Werthern. *Courtesy of the Center for Dubuque History, Loras College.*

> *some of the farmers abandoning their fields, already put in crops, for the uncertain prospects held out in the gold diggings. The same influences which existed in 1849 obtained this season also, without variation.*[7]

While this statement attests to the widespread thinking that lead mining all but ceased with the mania to "get my share of the rocks" in California, it is certainly true that lead mining persisted in the region for many years, even into the twentieth century, though at continually declining rates from the mid-century levels.[8]

Production of lead in the region had been slowly growing in the 1820s but sharply increased beginning in 1833 and peaked in the years 1845, 1846 and 1847. The following year, 1848, proved to be the harbinger of the continuing decline in production yet to follow.

LEAD PRODUCTION 1840–1859[9]

1840	13,425	1850	19,734
1841	15,988	1851	16,594
1842	15,939	1852	14,319
1843	19,574	1853	15,700
1844	21,864	1854	14,827
1845	27,248	1855	15,063

1846	25,884	1856	15,248
1847	27,134	1857	17,082
1848	23,869	1858	15,600
1849	22,109	1859	14,000

But lead mining was not yet a total bust, as we see in a letter from "Ella," copied from the *Chicago Dollar Newspaper* and printed in the *Miners' Express* in December 1849: "The great staple article of this part of the world seems to be lead. Every few days the news is brought to town [that] a new lead has been struck, and some poor laborer is hurried into the possession of wealth beyond his most dazzling dream."

The local economy was heavily dependent on lead-mining activities throughout most of the 1840s, and the Dubuque miners had reason to continue to have confidence in the lead trade. In September 1848, the steamer *Dubuque* freighted to St. Louis 6,477 pigs of lead, at 75 pounds each, for a total shipment of 243 tons and a value of almost $18,000. At that time, almost every vessel heading downriver carried some cargo of lead. The price of lead in St. Louis continued strongly in 1849, fluctuating between $3.75 and $4.25 for that twelve-month period.[10]

Further, additional records providing tonnage and dollar value indicate that the lead industry continued strongly in Dubuque into the 1850s.

Lead Shipped from Dubuque 1852–1855[11]

1852	4,234 tons	$348,000 value
1853	4,380	$481,000
1854	4,385	$526,000
1855	5,262	$521,400

Nevertheless, the Upper Mississippi Lead District had seen its best days. In the years following 1847, lead-mining activity continued its slow decline, mainly due to two facts: the easiest obtainable lead had already been mined, some of the shafts being down to water levels; and the declining prices for lead.

Despite the relative youth of Dubuque in 1849, the basis for its economy beyond that of lead mining had been expanding and diversifying, and it was becoming a prosperous, growing community with bright prospects. The dollar value of imports greatly exceeded that of exports, especially in 1851 when the ratio was 5:1. In the next four years, however, that number declined to a range of about 2.5–3.0:1.[12]

Dubuque in 1846. J.C. Wild. *Courtesy of the Center for Dubuque History, Loras College.*

In examining population growth for the city, county and state, one sees that the number of immigrants substantially outpaced the number of residents who left Dubuque for greener pastures. This was the time of the great government land sales at $1.25 per acre, and this was for *Iowa* land, some of the best on earth.

The resident population of Dubuque County doubled from some six thousand at the time Iowa achieved statehood in 1846 to over twelve thousand in the short period of six years up to 1852. The number of immigrants passing through this frontier river city to points in the near and far West could easily have been another ten to fifteen thousand. All of these would have been searching for the goods and services supplied by the merchants and businesses of Dubuque, leading to a dynamic and rapidly increasing local economy.

A contemporary account written in 1851 provides ample evidence of an attitude encountered by immigrants with the following xenophobic description:

> *A number of Prussian emigrants of the better class, with their beards, good figures, & foreign customs, a party of Irishmen, said to be "noble," a certain*

> *officer of the army undoubtedly "royal," who amused and astonished us by his wit & extensive information, merchants from St. Louis & the east, & Wisconsin & every state in the Union, with Canada & Europe, were found in the Cabin. On deck were Germans and Irish, a filthy set, whose uncleanliness no doubt hastened the deaths which occurred among them, & I was heartily glad when we landed the last at Dubuque.*[13]

As the influx of new arrivals poured into Dubuque County, those born within the state continued to be outnumbered by those born elsewhere: 36 percent had been born in other states, and 40 percent were foreign born, while 24 percent had been born in Iowa. It soon became clear that the Irish and the Germans made up a larger portion of Dubuque's population than any other group. Thomas Auge has determined from census records that in 1850 the population of the city of Dubuque was made up of 41.4 percent born in the United States and 58.6 percent foreign born. The Irish comprised 24.6 percent and the Germans 17.8 percent of the foreign born.[14]

Population growth in Dubuque city and county, as well as statewide, showed a steady increase for these mid-century years. The data for Dubuque County and the state of Iowa show this consistent growth for each of the years. The record for Dubuque city population, however, is not as reliable as for the county and state. Data for some years are lacking, while others show only estimates. Dubuque County, with firm figures, showed a continuing increase during the period. The city of Dubuque was then the commercial hub for the Iowa region, and one must conclude that its population increased apace with that of the county and state.

Population figures for 1846–1852[15]

Year	City	County	State
1846	n/a	6,030	102,388
1847	n/a	7,440	116,454
1848	4,000 est.		
1849	3,500 est.	9,185	154,573
1850	3,108 (4,071 est.)	10,841	192,214
1851	6,000 est.	11,000	205,335
1852	7,000 est.	12,508	229,932

Dubuque buildings. Enlarged segment from Meyer's *Dubuque im Staate Iowa*, circa 1850. *Author's collection.*

By mid-century, in contrast to the earlier untamed frontier atmosphere of the 1830s and 1840s, the small community was beginning to take on an air of domesticity. At this time, there were 2,002 families living in the city in 1,952 dwellings, most of them wood-framed.[16]

Outside the city, agriculture dominated. The proportion of the economy devoted to agriculture was quite substantial throughout the United States in the nineteenth century, and the Dubuque area was no exception. Farming was an occupation that was familiar to almost everyone. Acquisition of government land was then relatively cheap, and farming the rich Iowa land seemed an attractive life for many European immigrants. Indeed, many of the lead miners in the Dubuque area also engaged in farming, even as their primary occupation, with mining activities taking place only as a sideline. By mid-century, many of these farmers, with the easiest-to-retrieve lead having been raised, largely abandoned the pursuit of mining but maintained their concentration on farming.

In terms of manufacturing, it only made sense that local firms took the raw materials that were available locally and turned them into finished products. As an example, John Kennet established, in 1848, a shot tower in the Upper Mississippi area, took the region's lead and turned it into lead shot. Dubuque's shot tower was erected in 1856.[17]

A church and a school were two factors seen as a sign of a civilized community and were highly desirable to a family seeking a place to settle. In Dubuque, the church presence arrived early on. The Catholic Church had established Dubuque as the location for its newly formed diocese in 1837, and its first bishop, Mathias Loras, arrived in the spring of 1839. There were, by mid-century in Dubuque, thirteen preachers and a diversity of religious persuasions: Episcopalian, Catholic, Methodist, Congregational, Christian,

Baptist, German Congregational and German Methodist. A Lutheran presence was established informally in 1853 and formally in the following year.

Where schools were concerned, Dubuque was determined that its young people would have the advantages of formal education. Two brick schoolhouses were erected in 1850, but teachers were generally poorly qualified and wages were low—$1.00 to $1.50 per week for women and $10.00 to $15.00 per month for men.

At the higher education level, the beginnings of three primarily undergraduate colleges were established, though today with these new names: Loras College in 1839, Clarke University in 1843 and the University of Dubuque in 1852. The Reverend Samuel Mazzuchelli, OP, opened a college on Sinsinawa Mound, just across the Mississippi in Wisconsin, about the year 1844.

The matter of health was a constant concern on the frontier, where seemingly minor health issues could quickly become life threatening. The presence of medical doctors, regardless of the level of their expertise, provided a certain sense of comfort to the early pioneers. To this end, there were eleven physicians and one dentist in the city in 1850. The local citizens had talked of creating a hospital, spurred on by an outbreak of smallpox in 1845 and cholera in 1850, but without result. Finally, in 1852, they achieved success.

A system of laws and courts to adjudicate disputes was one of the early needs of the community of lead miners and early pioneers in Dubuque. The local citizens had already built a substantial courthouse of brick to meet this need in 1839. By 1849, there were seventeen lawyers practicing law, outnumbering the thirteen who preached the gospel. Apparently there was a need for this contingent of lawyers as it was reported that "the principle [*sic*] amusement of the people seems to be playing cards, Sunday and all. The law they carry in their pockets, and are ready to read a chapter on the slightest occasion."[18]

The city was definitely on the cusp of a tremendous growth spurt, and optimism ruled the day. In 1849, there were eight or ten hotels in Dubuque, the number depending on one's definition of hotel. There must have been a large transient population to take advantage of these facilities or they would not have existed. New immigrants planning to settle in Dubuque also needed temporary quarters while making more permanent arrangements.

All the business enterprises, as well as people contemplating going into business and private individuals, needed banking services. To meet these needs, local citizens incorporated the Miners Bank of Dubuque in 1836. It closed its doors in 1849, but by then, a handful of new banks had taken its place.

Julien Hotel, built in 1844 and augmented in 1854. Seen here as it appeared circa 1880. *Courtesy of the Center for Dubuque History, Loras College.*

At this time (1849–50), newspapers were the primary means of mass communication, there being no radio, television, telephone, Internet or Twitter. Dubuquers had access to six local newspapers, an apparent abundance for a community of only four thousand and a county population of some ten thousand.

As for transportation, Dubuque had at its doorstep the four-lane highway of its day—the mighty Mississippi. The first steam vessel to make it to the Dubuque area came in 1823, and regular traffic was established in the 1830s. By the time of the California gold rush, river traffic had reached a level of frequency and reliability that made it one of the keys to local economic development and prosperity.

Central to much of the steamboating through this period was Captain Daniel Smith Harris commencing in 1829 and continuing until the sinking of his *Grey Eagle* in 1861. Harris must have been viewed in awe if Josiah Conzett is to be believed: "As for the pilots of those days, well, they were the whole thing. When they swaggered uptown, hats on side and nose in the air, why, it was a distinction to be noticed by them; and to be cussed by them was a never to be forgotten honor."[19]

The extension of a number of streets to the river had not yet been accomplished by 1850. And in the matter of uptown streets, a local wag

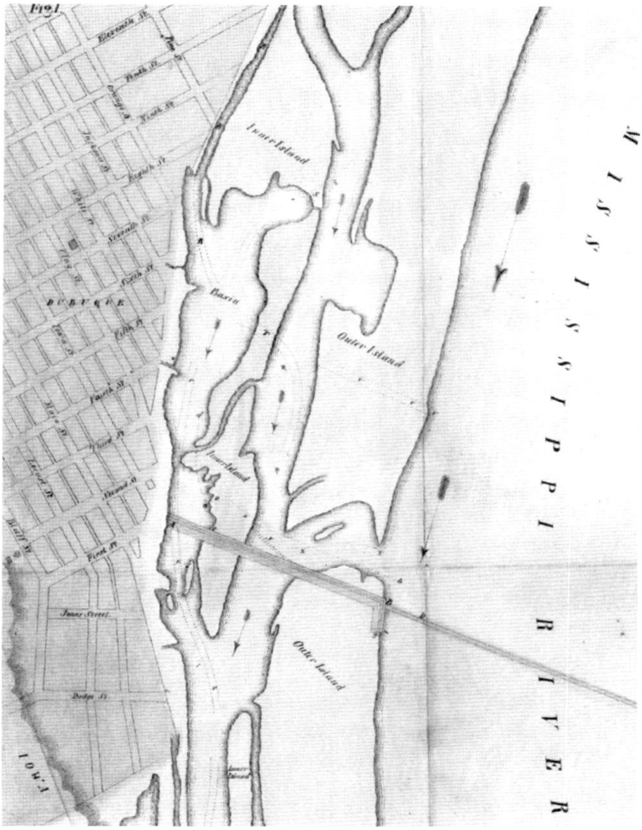

Left: Streets leading to the Mississippi River at Dubuque, 1855. *Author's collection.*

Below: Proposed causey providing Mississippi River access at Dubuque, 1855. *Author's collection.*

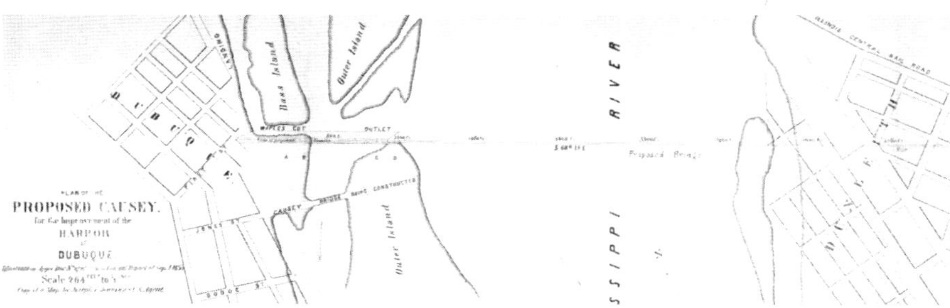

described Lorimier Hollow Road, leading to Dubuque's upper elevation, as follows: "Several serious accidents have recently occurred on this much-traveled, very crooked, pinched-up, starved-out, narrow-contracted, Lorimier-fenced-up, disreputable, dangerous, leg-breaking, skull-cracking, wagon-smashing, horse-killing, badly located, poorly-worked, corporation-neglected, tiresome and hilly road."[20]

Regarding pathways outside Dubuque, in 1839, Lyman Dillon plowed the furrow that became the corridor for the Military Road, providing a path for migrants from Dubuque to Iowa City. But even before this, the earliest settlers had already established other roads. They did this as a matter of practicality; people were moving from point to point on land. As the population in eastern Iowa increased, what had been the main "highway," the Mississippi River, was gradually being supplemented by roads inland.

Hope was held out for some years for extending the Galena and Chicago Railroad the few additional miles to Dubuque, but the road did not even arrive at Galena until late in 1854. It wasn't until 1855 that the railroad finally achieved termination on the eastern shore of the Mississippi at Dunleith (now East Dubuque, Illinois) opposite Dubuque.

Much agitation arose at mid-century concerning the development of rail lines, sometimes accompanied by overly optimistic and grandiose thinking. Given the threat the railroads presented to Galena (Dubuque's nearby economic competitor), with its commitment to river commerce, it is not surprising to see one of its newspaper editors mockingly refer to a projected rail line of its up-and-coming rival Dubuque as "the Dubuque, Pacific, Japan, and Shanghai Railroad."[21]

Lower rapids at Keokuk, Iowa. *Hall and Whitney.* Report of the Geological Survey of the State of Iowa. *1858.*

During this time in the 1850s, the Upper Mississippi Valley was about to undergo a substantial directional change in its commercial relationships in the region. In the early days of Dubuque, its outlook was downriver to its main trading partner, St. Louis, and points to the south. The Mississippi River, however, was not always reliable, being at times frozen over and at other times at low levels, bringing the problems related to two sets

Dubuque im Staate Iowa, circa 1850. *Author's collection.*

of rapids into play. But now, with the prospect of railroads coming into the area, a new directional perspective was coming to the fore—to Chicago and points to the east.

Thus, the battle lines were drawn between the slow, relatively cheap transport via the Mississippi River and the quicker, year-round but more expensive transport by railroad. The river traffic was, of course, confined to where the river went, while the railroads could lay track pretty much wherever demand called for it. The war did not last long, with Chicago and the railroads the victors. And so began the decline of the Mississippi River as the main avenue of commerce for the growing city of Dubuque and the gradual rise of Chicago and points east as the new direction of trade.

And so, we have now examined some of the factors contributing to a certain level of comfort for the citizens of Dubuque. We have identified some of the conditions that would make its residents want to settle down there, but for some, make them want to move on. A concise, picture of the commercial services then serving the residents was provided by the *Miners' Express*:

> *Twenty-two stores containing a general assortment of dry goods, clothing, groceries &c., besides one wholesale and retail hardware store, two drug stores, two auction and commission merchants; one bookstore; eight fruit and provision stores; one boot & shoe store; two tinware manufactories; three bakeries; one manufactory of confectionery; seven master builders; six plasterers; seven master brick-layers; four painters; two master stone masons;*

> *one marble worker; two saddle and harness manufactories; seven boot and shoe manufactories; nine tailor shops; three milliners and dressmakers; three jewelers and watch makers; six cabinet and chair manufactories; five blacksmith shops; three carriage and wagon manufactories; two gunsmiths; three cooper shops; five butcheries; one soap and can factory* [and] *two livery stables. Two flouring mills had at this time been built, and were then capable of manufacturing two hundred barrels of flour per day.*[22]

The growth and development of Dubuque at mid-century had progressed to a point where it provided a comfort level quite adequate for the day. Its business and commercial aspects were well established, and vigorous future growth was expected by all. The rule of law had been imposed on its citizenry, and religious and cultural institutions were in place. In short, the city of Dubuque was quite a prosperous community, one in which the early pioneers would happily settle and live the good life.

It would have taken a strong attraction to energize them to move on. One such lure that enticed many was the availability of cheap land and the prospect of putting down roots on a farm that one owned and being able to continue in the agricultural life with which so many were already familiar. But the city exhibited numerous desirable aspects, and for many, Dubuque became their home. Yet there was to come an even stronger attraction—one much more difficult to resist.

Chapter 2

"Zounds! Ezra, There Be Gold in California"

James Marshall made the colossal discovery that would change the face of the young nation on January 24, 1848: GOLD! The location was at Sutter's Mill on the American River, near present-day Coloma (140 miles northeast of San Francisco and only 50 miles northeast of Sacramento). That discovery set the stage for one of the largest mass migrations in United States history. The allure of gold and the prospect of easy riches proved to be irresistible for people not only from North America but also from around the world.

California had been annexed to the United States under the terms of the Treaty of Guadalupe-Hidalgo of February 2, 1848, concluding the Mexican-American War.[23] Neither party to the treaty knew of the vast treasure secreted in the soils of California, having signed the treaty only a few days prior to the discovery of gold at Sutter's Mill. Who knows what would have occurred had Mexico been aware of the enormous riches in the lands it ceded to the United States?

The news of the discovery spread around the planet, though the speed of travel varied tremendously. Sailing ships carried the news to the far corners of the globe, but given their commercial schedules and ports of call, China received the news prior to its arrival in New York, and London and Australia heard the news only a short time apart. Hawaii had the news in June, as did Mexico, and a shipload of gold seekers left Peru on November 30, 1848. Gold seekers came not only from the United States but also from Europe, South America, the Orient and even far-off Australia and New Zealand. All were drawn by the perennial and universal fascination with this precious metal.

Communication between California and the United States involved a significant time lag in 1848. The initial publication of the discovery of gold appeared in the *Californian* (then located in San Francisco) on March 15, 1848: "In the newly made raceway of the saw mill recently erected by Captain Sutter, on the American fork, gold has been found in considerable quantities. One person brought thirty dollars worth to New Helvetia [Sacramento], gathered there in a short time…Gold has been found in almost every part of the country."[24]

On April 1, 1848, the *California Star* (San Francisco) published its "Great Express Extra" and made special arrangements for "express" delivery of two thousand copies to Missouri. They were to arrive months later in August of that same year. "The Great Sacramento Valley has a mine of gold…From all accounts it is immensely rich and already we learn the gold from it, collected at random and without trouble, has become an article of trade…This precious metal abounds in this country."[25]

The report of the gold discovery arrived in Washington, D.C., on August 2, 1848, in letters provided to the president by a merchant of Monterey, Thomas O. Larkin, who was also a confidential agent for the president. President Polk had also obtained information from the report of Colonel Richard Barnes Mason, dated August 17, 1848, at headquarters of the Tenth Military Department at Monterey. These documents convinced the president of the authenticity of the gold discoveries in California and were printed as part of the president's State of the Union *Message*, as were two additional reports from Larkin. Polk stated, "The accounts of the abundance of gold in that territory are of such extraordinary character as would scarcely command belief were they not corroborated by authentic reports of officers in the public service."[26]

It seems clear that this announcement, coming from the president of the United States, was a significant factor in erasing the natural doubts that would have existed among the public concerning the amazing claims being made. These reports, of course, received wide distribution through the newspapers of the day and tended to confirm that the journey to California in search of gold would not be a fool's errand.

In the States, many of the initial newspaper reports were cautious and even dubious in presenting news of the gold discovery. But after Polk's comments and the not-to-be-doubted fourteen pounds of gold delivered to Washington, D.C., to accompany Mason's reports, the news created the widespread fever that turned into the gold rush, a truly transformative event that shaped the land into a coast-to-coast nation.

President, James Knox Polk. *From* The National Cyclopaedia of American Biography *(New York: T. White, 1896), vol. 6, 265.*

As word of the gold discovery spread, it inevitably was to arrive at Dubuque. We know it appeared in St. Louis in August 1848. One can conclude, given the flow of commerce between Dubuque and that city, that word-of-mouth news of the gold discovery would have appeared shortly thereafter upriver at Dubuque, in as little as ten days. Additionally, newspapers and letters from friends in the East would have provided sufficient information to raise the curiosity, if not the avarice, frequently attendant to the topic of gold.

Surprisingly, the first notice in Dubuque of the California gold discovery appeared two months later in the *Miners' Express* on October 4, 1848, in a piece reprinted from the *New York Sun.* The article tells of a gentleman who had brought in a sample of "gold." The bottom line, however, states that he only provided another proof that all that glitters is not gold, as the sample was nothing more or less than species mica.

The choice of this article to be reprinted in the *Miners' Express* indicates that the local editor concurred with the theme presented: don't believe all you hear of rumors and reports of gold discovered in California. His caution appears to have been founded on the dictum, "If it sounds too good to be true, it is probably not true."

The first truly substantive notice indicating there was some merit in the reports of California gold appeared in the *Dubuque Tribune* on October 20, 1848. Under the headline "Gold in California," it printed the Larkin letter in full. He had personally visited the gold regions and reported, "It is all as represented" and there were vast quantities of gold.

The next newspaper notice of the gold discovery appears in the *Miners' Express* in a short article reprinted from the *Louisville Courier* noting that a group was forming at that place to go to the gold fields. It appears without editorial comment. Then, on December 1, we read of the capture of naval deserters from the ship *Warren.* These men were heading for the gold fields but were captured and returned to Monterey to be court-martialed. A few days later, another brief notice reports that the steamer *Sarance* was departing Cincinnati loaded with ploughs and plantation wagons for California in order to make splendid farms where there "tis gold dust in plenty." In addition, the news got closer to home with a notice that a public meeting was to be held on December 9 in nearby Galena for those wanting to go to California.

The people in this region of the Upper Mississippi Valley were evidently well aware of the great find in California, but the news of this public meeting was the first newspaper account tying the gold discovery to the local citizenry. A few additional newspaper articles of a hesitant, if not starkly negative, tone appeared in the following weeks. They are characterized by phrases like "sacrifice health," "lose friends, fortunes, and character," "boundless prodigality," "endure sternest rigors of want" and "nameless grave." These articles, nevertheless, do report that the gold discovery was causing great excitement and the gold fever was "raging."

Doubts about the abundance and ready availability of the gold were largely dispelled with the publication of President Polk's *Message* in the *Miners' Express* in two installments on December 19 and 26, 1848. In the first of these, the news appeared that is commonly credited with being the catalyst that caused the gold mania to erupt: "Explorations already made warrant the belief that the supply is very large and that gold is found in various points in an extensive district of country."

Even with this report, from the president of the United States no less, the local editor hesitated, and among his accompanying editorial comments there appears mention neither of California nor the gold, though he does comment on other items in the *Message.* Finally, in the next week, in the editor's comments following the concluding installment of the president's *Message*, we read:

> *We have been slow to credit the reports which have from time to time reached us of the astonishing fertility of the Gold Mines of California. Some few weeks since, we published what we then thought was the most probable report from that country, namely: that the substance supposed to be gold was nothing more than "pyrites." The reports which have subsequently reached us assume more of an authentic character, and are calculated to inspire the belief that, what seemed to be the wild fabrications of some visionary adventurer are real, substantial facts.*[27]

And so it seems evident that the magnitude of the California gold discovery reported by the president of the United States himself, and the box of gold delivered to Washington, D.C., by government officials, together provided the not-to-be-doubted evidence that the newspaper editors of the country needed to convince the populace that the stories were true. There was indeed a huge amount of gold in California, and just as good, it was there for the taking, California not yet having achieved statehood and without a system in place for imposing levies on gold discovered by the miners. It didn't take long for tens of thousands to be infused with and inspired by the French phrase *laissez les bons temps rouler* (let the good times roll).

Chapter 3

Scratch Where It Itches

The excitement caused by the breaking news of the ready availability of gold in California called for action, and the early pioneers in the Upper Mississippi Valley were nothing if not men of action. Men discussing with other men the possibility of getting rich in the gold fields would have reinforced one another in the desirability of taking advantage of a never-to-be repeated opportunity. Men discussing this opportunity with their wives and families encountered ideas intended to dampen their enthusiasm, ideas that led them to consider the effect on family, the hazards to be encountered and the costs to be borne.

It would be difficult to deny that many, if not most, of the Dubuque forty-niners were invigorated by the idea of getting rich, with a goodly portion of the early travelers adding the modifier "quickly." Among them were those who probably intended to continue the good life in far-off California, but it is likely that many intended to return back east to be among friends and to use their newfound wealth to provide the good life for the families they had left behind.

The average wage for Dubuque laborers at this time was about one dollar per day. Compare that, and the lead miners surely did, with the newspaper reports of those who struck it rich in the gold fields, where ordinary recompense was expected to be twelve to fifteen dollars per day. Small wonder that the miners and others in the Upper Mississippi Lead District were taken by the prospect of sharing in that wealth.

The California gold excitement evidently reached a fever pitch in Dubuque County in December 1848. On the twenty-ninth of that month, Dubuque entrepreneur Richard Bonson recorded "tremendous excitement

MAIN STREET, DOWN TOWN, DUBUQUE, IOWA.

Dubuque Downtown and Uptown, 1857. *From* Ballou's Pictorial Drawing Room Companion*, October 31, 1857. Author's collection.*

MAIN STREET, UP TOWN, DUBUQUE, IOWA.

Richard Bonson. *Courtesy of the Center for Dubuque History, Loras College.*

Reverend Joseph Crétin. *Courtesy of the Center for Dubuque History, Loras College.*

about Callifronia gold mins—many people talks of going thear."[28] The following day, on December 30, a public meeting was held at the courthouse in Dubuque to form an organization of all who had an interest in going to the gold fields. Among the first to leave was William H. Merritt, then proprietor of the *Miners' Express*, along with Edward Mobley, S.M. Hammonds and J. McCoy, who departed on January 1, 1849, to travel to California via the Isthmus of Panama.

In April of that year, the *Miners' Express* reported the names of over sixty citizens of Dubuque who had gone to California.[29] Bonson himself had a touch of the fever, as he "bought a Callafronia wagon and 4 yokes of oxen for 160 dollars." This may, however, have been just a business investment, as he himself never joined the forty-niners. He may have been planning to send a proxy, as he records in early 1850 that he "almost concluded to send Wm. Delaney & others with oxen to Callifronia."[30]

Additionally, and somewhat surprisingly, a few of Catholic Bishop Loras's small number of priests were also taken with the urge. The Reverend Joseph Crétin, later to become the first bishop of St. Paul, the Reverend J.G. Perrodin, the Reverend André Trevis and the Reverend John McEvoy all left a written record of their interest. Their bishop was not enthused—permission denied.[31]

Their thinking was not that much different from the rationale that brought the Cornish immigrants who had settled near Dubuque at Mineral Point in Wisconsin to work the lead mines there. "Their first intention in coming to America, getting rich and returning to Cornwall, was not realized; and though they did not get rich, they considered the conditions into which they had come far better than those they had left."[32]

There was also at that time, just as now, the natural restlessness of a younger population. Many of these folks were already familiar with migration, life "on the road" and the severe rigors and hardships encountered. It must be remembered that most of the Dubuque population had already become familiar with the difficulties of migration, as almost all of them had immigrated to Dubuque subsequent to the 1833 opening of the territory, a scant fifteen years prior for the earliest arrivals and even fewer years for those who arrived in the late 1830s or even in the 1840s. Given the steady tide of population movement from the eastern seaboard states to the more western-situated states, it became almost the natural thing to do. "Everyone's doing it." If you had moved once, the second move became easier.

Partly, too, it was the "grass is greener" syndrome that brought many pioneers to Iowa. And with government land to be had at $1.25 per acre, the chance to have their own property proved to be an irresistible lure for many an enterprising individual or couple. This can be seen in the rationale of the Langworthys' interest in coming to the Dubuque area in an earlier time. Lucius and Edward Langworthy, "being curious to see the country," obtained their father's consent to make this journey in 1827 "with stout hearts." Their sisters, Mary Ann and Maria, who accompanied them, were, on the other hand, noticeably less enthused, being described as "pale with sorrow" at the prospect.[33]

In the 1840s, great sentiment regarding Manifest Destiny had been instilled in the populace via the newspapers of the time in their coverage of the federal government's interest in the lands of the West. The Oregon Question was then prominent, and the Mexican-American War, concluded in 1848, brought present-day Arizona, Utah, Nevada, California and parts of Colorado and New Mexico into the political orbit of the United States of America. These events helped inflame the passions of those who could envision some prospective self-interest in these vast territories being brought "into the fold."

> *There have been, from time to time, rumors of a better country to the west of us, and a sort of a pioneer, or western fever would break out among*

> *the people occasionally. Thus in 1845 I had a slight touch of the disease on account of the stories they told us about Oregon. It was reported that the Government would give a man a good farm if he would go and settle, and make some specified improvement.*[34]

Beyond the lure of riches and the natural restlessness of the times, there were other reasons for leaving Dubuque and heading west. The dreaded cholera was one of them. Citizens everywhere in the United States, but especially those who had settled anywhere along the Mississippi River, were acutely aware of the scourge of cholera, an infectious bacterial disease that often brought terminally disastrous consequences. The disease was acquired through the ingestion of contaminated food or the drinking of contaminated water. Onset of the illness occured within one to five days after the initial infection. It was marked by severe, watery diarrhea and accompanied by nausea and vomiting, muscle cramps and rapid onset of dehydration, leading to shock and, frequently, death. All of these could occur within the course of a single day.

The most readily available and effective treatment at that time was rapid rehydration—drinking lots of water. But since the organism causing the disease was often found in water, at times this must have only made the problem worse. That the treatment was so simple was likely not widely known. For those selected for death, it was often a matter of resigning themselves to their fate, having drawn the "short straw."

Richard Bonson took note of the presence of cholera on the Mississippi, entering in his diary, "The collara is very bad in St. Lewis—many died on it." As for Dubuque, "hear of some collara in town—Lord help me." A week later, it merited another entry: "Hear that people is dying of collara—Mr. Lewis is sick—like to die," and on the following day, Independence Day 1849, "hear that Pat Thornton died of collara today—the collara is very severe."[35]

Other possible reasons for leaving Dubuque to go to California were likewise negative, if also having the intention of improving one's life. There were undoubtedly those who left in order to escape debts, those who may have found themselves in unhappy marriages and those who did not rank highly in the family pecking order (i.e., not the first-born son). Poverty, escaping contracts or other obligations and fleeing from justice can also be presumed to have been the rationale for some.

Despite whatever enthusiasm fueled their fervor, these prospective forty-niners would quickly have encountered the reality of cost. In assessing the expenses of the long, arduous trek to California, it will be useful to know something about the wages available to those laboring in Dubuque and the mines at the time. To say that the overland trip cost, for example, $300 tells us little unless the value of that $300 is known. A worker in the Wisconsin mines in the 1840s was earning $1.00 a day, and this would seem to be a pretty good guide for our purposes here. Using another approach, there are conversion tables available that convert values of past years to the value of dollars in the current year. From these it is determined that $1.00 in 1850 equates to approximately $28.80 in 2010 dollars. The average laborer in Dubuque would have been making about $1.00 per day and working six days a week, for an annual average salary of about $300.[36]

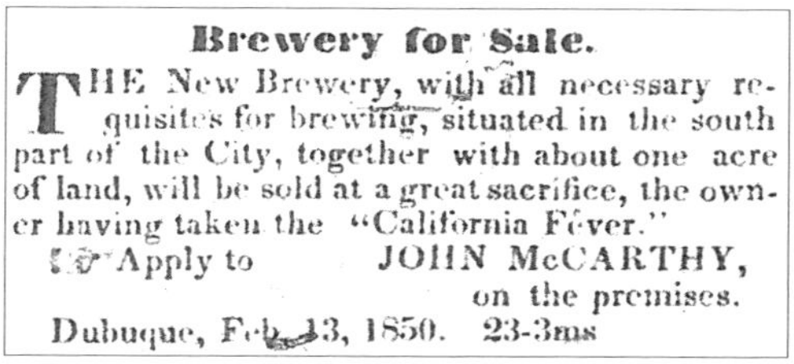
Brewery for Sale.

THE New Brewery, with all necessary requisites for brewing, situated in the south part of the City, together with about one acre of land, will be sold at a great sacrifice, the owner having taken the "California Fever."

☞ Apply to JOHN McCARTHY, on the premises.

Dubuque, Feb. 13, 1850. 23-3ms

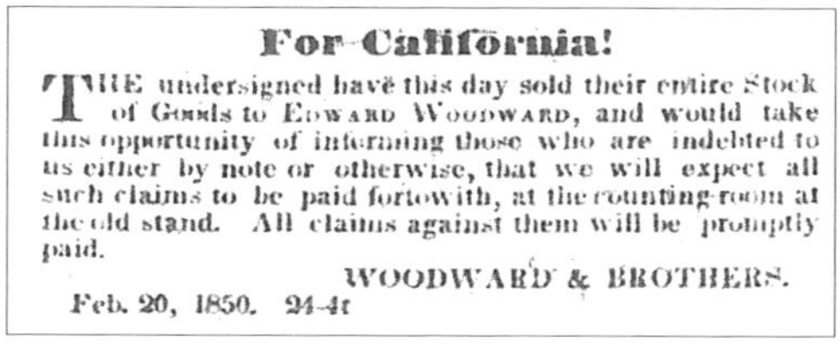
For California!

THE undersigned have this day sold their entire Stock of Goods to EDWARD WOODWARD, and would take this opportunity of informing those who are indebted to us either by note or otherwise, that we will expect all such claims to be paid forthwith, at the counting-room at the old stand. All claims against them will be promptly paid.

WOODWARD & BROTHERS.

Feb. 20, 1850. 24-4t

"For California" and "Brewery for Sale." *Miners' Express*, 1850.

Those leaving for California from Dubuque on the eastern border of the great Trans-Mississippi West had three basic options open to them: 1) overland via the Mormon/Oregon Trail (also to become known as the California Trail); 2) down the Mississippi, across the Isthmus of Panama and by sea-going vessel up the Pacific coast to San Francisco; and 3) the maritime journey around "The Horn," the southern tip of South America, and thence up the coast to San Francisco. Cost, time of the season, time of the journey and danger to be encountered were all factors in determining which of these would be selected.

If one were truly smitten by the gold bug and just *had* to leave immediately upon reading President Polk's confirming *Message* in the Dubuque newspapers, that would have meant leaving during the winter months. Since the winter season was not at all practical for an overland journey, the choices were the all-sea route around "The Horn" or the sea/land route across the Isthmus of Panama. The latter seemed preferable and was, in fact, chosen by a number of Dubuquers. William H. Merritt and his companions, among the first to leave Dubuque on January 1, 1849, headed downriver to New Orleans for California via the Isthmus.

Dubuque forty-niners leaving later in the season found the estimated initial cost for the overland trek to come to about $300. Among the items needed were mules, saddle, bridle, halters, ropes, blankets, rifle, pistols, knife, flour, bacon, sugar, coffee, tea, etc.[37] Over and above the cost of these items, the forty-niner had to be prepared for incidental costs such as ferry crossings, cattle feed and living costs while engaged in mining or other pursuits after arriving in the gold fields; this amounted to another $300. Dubuquer John Coffee recovered some of his costs by selling his wagon and oxen when he arrived in California, realizing $380, not a large amount in the California economy but enough to get him to the gold fields and tide him over until his search for the precious metal became productive.

The high costs of the trek west were brought home to Dubuque by an overlander from the small town of Ringwood Prairie in McHenry County south of Chicago. As he was passing through Dubuque, probably on March 28, 1850, he noted in his diary, "I found Dubuque a large place, 3,000 inhabitants, and every one wide awake to make something off the Californians."[38]

Others heading west and passing through Dubuque were likewise dismayed at the high prices encountered. A traveler from Lima in Rock County, Wisconsin, Miss Polly Coon bears witness. Her journey (to Oregon—not California) began with her parents in the Crandall Train on March 29, 1852. By April 10, they had reached the Mississippi River at what they identified as Eagle Ferry, some two miles above Dubuque, and crossed over on the eleventh. They left the city and passed through "the deepest mud and worst roads I ever saw." They then "camped in a field and got about half enough poor hay for which the man charged 30 cents per yoke. I record this as a demonstration of the depth of heartlessness to which the human heart is capable of arriving." It will not take too much imagination on the part of the reader to suspect that this "heartlessness" had already been experienced by many a forty-niner and was only a warm-up for what was yet to come on the overland trail.[39]

Family and friends received many letters from the Dubuque forty-niners about the excessive costs of the journey and in the gold fields, but they reveal nothing about the source of the money necessary to make the trip. Pooling of family resources, loans from friends or banks and "selling out" would have been among the methods used to raise the necessary funds.

No one knows the names of all the companies leaving Dubuque for California in 1849 and ensuing years, as some were just small groups of friends or family members and likely did not leave a historical record of the

CALIFORNIA SALES!

AS we intend starting for El Dorado in a few weeks, we will sell off the balance of our fine stock of READY-MADE CLOTHING *ten per cent. less than Eastern cost.* The stock is fine and fashionable, and those who want a splendid suit for spring, at almost nothing, will find this their chance.

Call at Bennett's store, next door to W. L. Johnson's Auction house. mar 2–tf

California Sales. *Dubuque Tribune,* 1849.

event. Often these small groups would join up with others at Kanesville (now Council Bluffs), Iowa, the jumping-off point for overlanders from Iowa.

The size of the Dubuque companies varies from as small as five or six people, such as McGarvine's and Shipton's, while Kennedy's Company was composed of only four wagons. Then there were the larger companies: the Galena and Dubuque Company of fifty-three teams, the Great Dubuque Company of thirty-two wagons and the Dubuque Emigrating Association of fifty-four men and twenty wagons.

Additionally, there were other teams leaving from Galena, Illinois; Andrew in Jackson County, Iowa; and other locales in the Dubuque area. Dubuquers could easily have joined companies leaving from these other communities, just as people from these outlying communities undoubtedly joined Dubuque companies.

The evidence for the names of these companies comes frequently from letters written by the Dubuque overlanders that were published in the Dubuque newspapers. It is not altogether clear when a letter writer refers, for example, to Gratiot's Company if he is referring to a company headed up by Gratiot or to a company headed by someone else, of which Gratiot is simply one of the members.

Some of the events and occurrences on the trail and in California were so prevalent in the lives of these Dubuque gold seekers, or so striking—as in "seeing the elephant," i.e., seeing something quite out of the ordinary—that they became natural topics for letters written back home. They often wrote descriptions of the difficulties of the journey. Life during the trip, whether by land or sea, was so arduous that the writers frequently included in their letters thoughts regarding the advisability of others making the trip and even questioning their own sanity for undertaking the journey.

Several Dubuquers clearly expressed their advice in the negative. The California trip was not worth the great deprivations suffered. One such rather stark narrative is supplied in a letter written by Dubuquer Edmund M. Allen to his friend J.C. Weatherby: "Just tell our friends that talk of coming across, that they had better stay at home and not try it unless they want to nearly starve, get the scurvy, the diarrhea, and every other kind of disease you can mention; nearly famished for water; hot weather, dry weather, and dust for miles so thick you can scarcely see."[40]

There were, nevertheless, many Iowans who were included in the gathering throng of forty-niners. One writer estimates that 1,200 from the state joined the gold rush. An additional unknown number from throughout the Upper Mississippi Lead District would also have left their footprints on the overland trail.

Chapter 4

GOIN' TO GIT MY GOLD

Of the three routes to the gold fields—overland, via the Isthmus of Panama or around Cape Horn of South America—most forty-niners chose the overland route, perhaps because it was the least expensive. But for others, this could mean trekking all the way from the East Coast of the United States and a much greater commitment of time and resources than for those leaving from states like Iowa or Missouri.

Traveling overland from Dubuque would have meant a journey of about 2,100 miles, averaging perhaps 15 miles per day, for a total of about 140 days of travel. Parts of the overland trail had previously been established in earlier years: the Oregon Trail and the Mormon Trail. The western extension was added in 1849, and the whole became known as the California Trail. It was open to travelers in the summer months, when grass for the livestock was plentiful and they could avoid freezing temperatures on the plains and massive snowfalls in the mountains.

For those who wanted to travel at other times of the year, they could head south to New Orleans. There, they could book passage on ocean-going vessels (steam and/or sail), cross the Caribbean Sea to Chagres on the Atlantic Coast of the Isthmus of Panama and thence overland to Panama City on the Pacific Ocean, where they would again board a marine vessel to San Francisco. One didn't have to do a lot of walking on this trip, but especially in 1849, there was an insufficient number of ships on the Pacific side to handle the large number of men gathered on the Isthmus, resulting in lengthy delays and rising prices in getting to San Francisco.

The all-sea route was a journey that could last from four to six months and often turned into an adventure never to be repeated. William Bothwell, writing to his friend Dr. Horr in Dubuque, narrates his terrible escapade:

> *The intensity of the cold, the fury of the wind, the lofty billows, and the heaving and tossing of our noble barque were well calculated to fit a person for believing "big yarns." We were 12 days at one time in a furious gale, and five at another, with helm lashed, the ship at the mercy of the winds, all sail in, the seas mountain high—some of which condescended to come down upon us and take passage in our state-rooms. On August 20th, six inches solid ice on the front part of the vessel; water frozen so that it would not run from our casks; thermometer 24 degrees below zero. To look around the cabin and see each man pent up without fire, and peeping out of his room at his fellow sufferers presented so much the appearance of culprits looking dolefully at each other through the grates of some state prison that I was constrained to believe that it would be expedient for the United States to abolish capital punishment and substitute a trip around the Horn, and it is my opinion they would never commit a second crime, knowing the penalty.*[41]

Despite the rigors and dangers of rounding The Horn, Bothwell survived and was situated at Jaw Break on the Feather River in California when he penned the above excerpt from his letter of September 10, 1849.

Each of the alternative methods of getting to the gold fields was fraught with hazard. The California Trail meant crossing the Great Plains, the Rocky Mountains,[42] the desert basin of Nevada and Utah and, finally, the Sierra Nevada Mountains, as well as encountering and dealing with the resistance of the Native Americans, who were contesting the advances of the white man into their lands. One historian estimates that thirty-two thousand pioneers crossed the plains in 1849, forty-four thousand in 1850 and fifty thousand in 1851.[43]

The typical timing of this journey meant jumping off from various points on the Missouri River about April 1 but not much later than May 15. An earlier start was not possible because the needed forage grass for the livestock was not yet up. Nor could the departure be much delayed in order that there would be sufficient time to pass through the Sierra Nevada Mountains prior to the arrival of the winter snows. The result was that in a period of some six weeks, very large cohorts were all on the trail at much the same time.

Thus, at Fort Kearney, Fort Laramie and other points on the trail, there could be a total of 20,000 or even 30,000 passing through in a short period of a few weeks. In June 1849, Dubuquer Josiah Heacock reported that "6,000 had already passed this place [Fort Laramie] on their way to California," and equal or larger numbers were yet to come.[44] The chaos must have been stunning. Picture these numbers, as recorded by Henry Stine up to July 5, 1850, at Fort Laramie: "33,171 men, 803 women, 1,094 children, 7,472 mules, 30,616 oxen, 22,742 horses, 5,270 cows, 8,998 wagons."

One of the forty-niners, a Mr. Gilmore, concluded that he had had enough after a journey of 140 miles and reversed course and headed home, meeting 1,125 wagons heading west. He provides a picture less than the glamorous, high adventure expected by some:

> *A large number of emigrants had died with the cholera, and the disease was still among them. In one encampment over night eight persons died... Upwards of two hundred emigrants must have died on the Plains—and sickness still among them...The cholera has made its appearance among several of the Indian tribes on the opposite side of the river, and a large number have died. It is said to be raging to an alarming extent among some of the tribes. Several have also died with the small pox.*[45]

The land and sea route via the Isthmus of Panama, in addition to the hazards of the sea, meant landing at Chagres, a quite uninviting village at that time, where these travelers initiated their jungle crossing into swamps filled with disease-ridden mosquitoes and poisonous snakes. To top it off, cholera and yellow fever visited many of them, sometimes with mortal effect.[46]

Further, demand for bookings for the journey was so high that many vessels were quickly brought out of retirement and made marginally seaworthy and took on passengers with little more than hope that the planned destination could be reached. One estimate has sixty-two thousand, a staggering number, arriving at the port of San Francisco from April 1849 to March 1850, though not all of these would have arrived from the Isthmus.

The all-sea route was accompanied by overcrowding on the ships, inadequate provisions, raging storms and the possibility of a journey lasting up to six months or more. We have seen above Bothwell's graphically vivid description of rounding The Horn. He was not pleased.

For Dubuquers traveling the overland route—and most did—the first leg of the trip required getting across Iowa to the jumping-off site at Kanesville/Council Bluffs, Iowa. At that time, the Military Road was in place running from Dubuque through Cascade, Anamosa and on to Iowa City. We know that many Dubuquers followed this road south out of Dubuque.

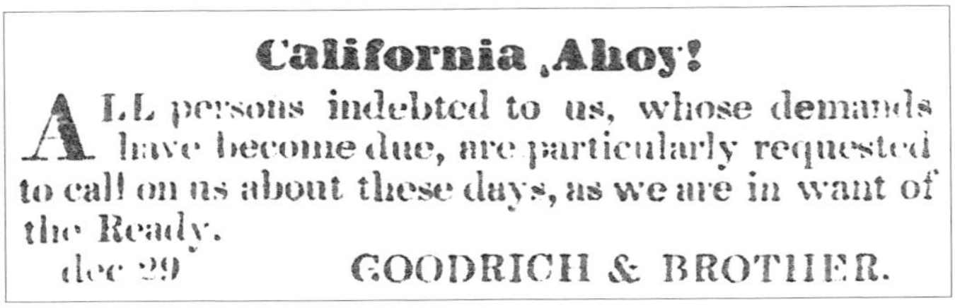
California ,Ahoy!

ALL persons indebted to us, whose demands have become due, are particularly requested to call on us about these days, as we are in want of the Ready.

dec 29 GOODRICH & BROTHER.

California Ahoy! *Dubuque Tribune*, 1848.

Beyond that, one route would have taken them through Marengo, Montezuma, Newton and Des Moines and from there to the jumping-off point at Kanesville. Alternatively, after leaving the Military Road at Iowa City, they could have connected to the Mormon Trail a little farther south. This road was pioneered in 1847 across the southern tier of Iowa counties in the Mormon remove from Nauvoo, Illinois, to the promised land in Utah. At least one Dubuque outfit took advantage of the Mormon Trail in Iowa. E.F. Gillespie, writing from near Kanesville to Thomas Hart Benton Jr., says: "We took a southern route within 12 miles of the Des Moines and crossed at Martin's Ferry, and intersected the Mormon Trail at Pisgah. This seems to have been the traveled road, and I suppose the best, as wagons which were before us that took the route by Fort Des Moines are still not here."[47]

The Oregon Trail and the Mormon Trail had been well established by 1849, but for Dubuquers who wanted to travel across the state to Kanesville, the route was less well known. In early February 1849, the *Miners' Express* helpfully provided narrative directions in time for those leaving in the spring of that year:

> *The following road through this State is the one that all emigrants from northern Illinois, Iowa, Wisconsin, and Michigan should take in order to secure good stopping places, ferries, and plenty of timber, and all other necessary requisites for the emigrants or their teams. The Distance will be found very nearly correct. From Andrew to Maquoketa seven miles, cross the Maquoketa on bridges; thence to the Wapsipinicon River, thro' South Grove* [?], *twenty miles; thence to Tipton, Cedar County, sixteen miles; thence to Washington ferry on Cedar River, ten miles; thence to Iowa City, fifteen miles...From the city cross Iowa River at the ferry; thence to Sigourney in Keokuk County, fifty miles, crossing English River at Wasson & Waters' mill; the country is thinly settled; thence to Eddyville on the Des Moines River, crossing the two forks of Skunk River at the bridges. From Eddyville twenty-five miles to Albia, Monroe County, some settlers, thence to Chariton Point, the center of Lucas County; here the large Mormon trail intersects the Eddyville road. This is one of the best roads in the*

State, twenty miles west of Chariton Point; here there are a few settlers. Thence to Pisgah, a Mormon town. Thence to Trader's Point on the Missouri River; here are four or five stores on the opposite side of the river, and an Indian agency.

This route has the advantage over the old ones in being some three hundred miles nearer, better timber, water, and every way adapted to the wants and wishes of the mover. A glance at the map will show this to be a due west course from Chicago, directly east of the South Pass.[48]

The Mormon Trail left "the states" near Kanesville crossing the Missouri River and tracking along the Platte River in Nebraska. It then proceeded along the Platte to Fort Kearney, on to Fort Laramie, across present-day Wyoming to Fort Bridger and then south into the Great Salt Lake Valley.[49]

There was general awareness of these trails among those contemplating the trip, but the forty-niners were eager for more detailed information.[50] Printed sources of various kinds were available, or soon would be, to meet the demands of a ready market thirsting for information about the Oregon Country and the West in general. Washington Irving, perhaps better known for other titles, was one of the early authors who both helped create the curiosity and satisfied it with three of his books: *A Tour on the Prairies* (1835), *Astoria* [Oregon] (1836) and *The Rocky Mountains* (1837).

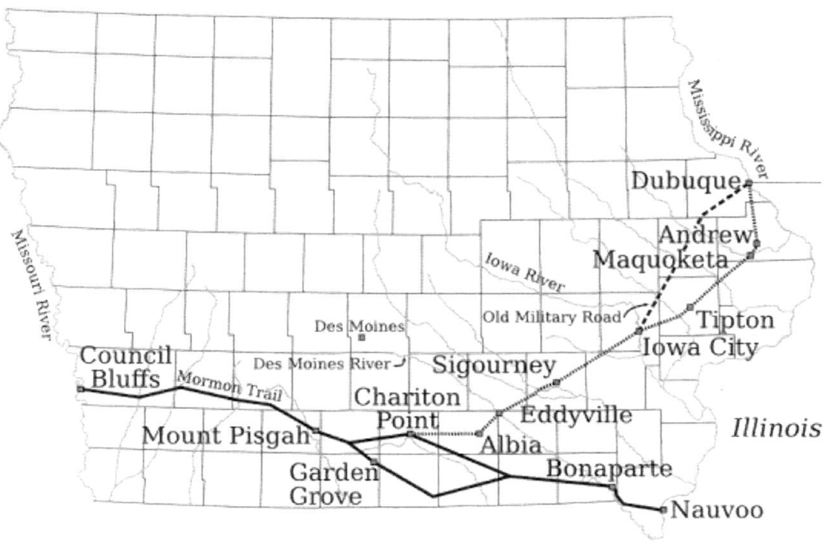

California Trail in Iowa. *Courtesy of Tom Klein.*

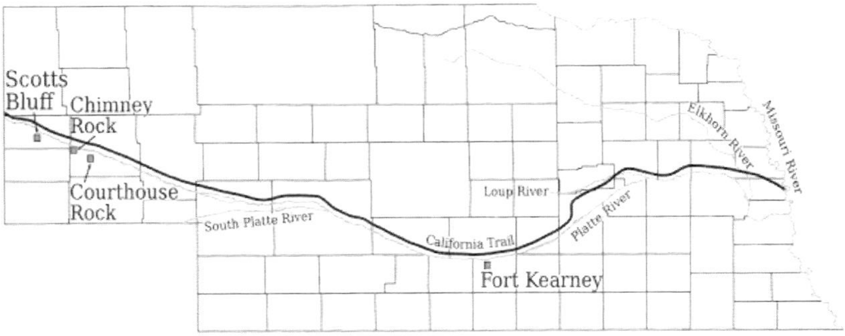

California Trail in Nebraska. *Courtesy of Tom Klein.*

Already, in 1842, William Lawther had established a bookstore in Dubuque, undoubtedly prepared to meet the needs of his clientele. Forty-niners from Dubuque could have acquired John B. Newhall's *New Map of Iowa with descriptive notes. Designed for the use of emigrants and travelers...* (1848). This guidebook would have gotten travelers across Iowa to the jumping-off point at Kanesville. In 1849, the *Dubuque Tribune* reported that printed items were available to inform prospective overlanders on what they were facing on their long trek and to prepare them for the journey.[51]

Washington Irving. *Author's collection.*

Then, too, there was Francis Parkman's *The California and Oregon Trail; being Sketches of Prairie and Rocky Mountain Life.* To meet the demand, this book went through three printings in March and April 1849. Sadly, it was neither a guide to get to California nor to Oregon, as Parkman had only traveled through the states of Nebraska, Wyoming, Colorado and Kansas. But it did serve to heighten general interest in the West. Publishing guidebooks was just another example of one of the peripheral activities designed to make a buck that arose around the gold rush.

At least twenty-five books or pamphlets specifically responding to the gold rushers were published in the years 1848 and 1849. Eighteen of them appeared in 1849, with two of them being in German. Everyone, including publishers, wanted to make a buck off the gold mania. The sad thing is that often these guides were a good bit removed from a strong sense of truthfulness. Distances were misrepresented on the short side, and the ease of the journey was overstated, as was the likelihood of achieving great wealth. Much was reprinted from other guidebooks, a practice well known in the newspapers of that time.[52]

In addition to these commercial publications, the federal government had been supporting a variety of explorations into its newly acquired lands for much of its early history. The various expeditions of John Charles Fremont were among these. While Fremont's reports, with maps, provided some guidance for a portion of the journey, they were clearly not a revelation as to the shortest or best trail to the gold fields; after all, that was not the purpose of these expeditions.

One of the most significant aspects of the Fremont reports is that they contain maps of the routes traveled with a level of accuracy and detail that previously had not been available, showing where water, wood and grass for livestock could be had. Significant also is the fact that the maps led the forty-niners relatively close to their desired destination: California.[53]

That the majority of those heading to California were relatively young and fit men will come as no surprise. On the other hand, there were also women who fell to the allure and excitement of the trek west or simply felt bound to follow their men. No less than the men, these adventurers were subject to the difficulties of the journey, as JoAnn Levy writes:

> *We spent three days very pleasantly although all were nearly starved for the want of wholesome food, but you know my stomach is not lined with pink satin. The bristles*

Francis Parkman. *From* The National Cyclopaedia of American Biography *(New York, T. White, 1893), vol. 1, 432.*

on the pork, the weevils in the rice and worms in the bread did not start me at all, but I grew fat upon it. Emily, Miss Bartlett and myself had a small room with scarce light enough to see the rats and spiders.[54]

Sickness and health were regular topics of conversation in the correspondence received from all travelers, whether Dubuquers or not. The recollection of one Juliet Brier sums up succinctly the long and short of illness on the trail: "Did I nurse the sick? Ah, there was little of that to do. I always did what I could for the poor fellows, but that wasn't much. When one grew sick he just lay down, weary like, and his life went out."

The presence of cholera in Dubuque in the years 1848, 1849 and 1850 was a matter of great concern and may have been one of the considerations some had in leaving for California. But whether traveling overland or by sea, there was no escaping it. The Dubuque forty-niners provide ample evidence.

William H. Merritt noted its presence at the Isthmus, describing cholera as "raging" in Chagres, with one hundred sick and eleven deaths in two days. As for those traveling overland in 1849, there was still a certain amount of naïveté regarding the pervasiveness of cholera on the trail. E.F. Gillespie, then near Council Bluffs, testifies to this fact in writing to his friend Thomas Hart Benton Jr. in Dubuque. He records the arrival of the *Highland Mary* with forty-five deaths on board from cholera. He then expresses what was a hope more than an informed opinion: "I think myself there will be no danger when we get out on the boundless prairies." How wrong he was.

John C. Fremont. *From* The National Cyclopaedia of American Biography *(New York, T. White, 1893), vol. 4, 270.*

Dubuquers, too, were counted among those taken away by the illness. An untimely death came to Valentine Glenat out in the middle of Wyoming. He had a short time from onset to death, followed with a quick burial by his traveling companions about 170 miles west of Fort Laramie on the north side of the Platte River. Thomas

Newcomer died on July 30 at Goose Creek, some 200 miles west of Salt Lake. He suffered for thirty hours and died "insensible of his suffering." Another, Dubuquer, R.W. Hart, also died from "inflammation of the bowels" (cholera) six days after arriving in California.

Despite the dreaded cholera, it was not even the most common disease to be feared. Scurvy, brought on by a vitamin C deficiency, was an equal or greater threat on the trail than cholera, taking away as many as ten thousand of the forty-niners.[55] One of its symptoms was darkened blood spots on the skin, often on the legs, resulting in it being referred to as "black leg." Dubuquer Josiah Heacock was a victim of scurvy. He was well aware of the cure but continued his suffering, as he was "not being where vegetables could be got." Aside from those trekking to California, these diseases also wreaked havoc on the Native Americans, who had not previously been exposed to them, killing untold thousands.

Death on the trail from a variety of other causes prematurely ended the California venture for a number of Dubuquers. Aside from the deaths due to cholera, James DePui was lost and presumed killed by Native Americans, as was Abraham Whitesides and a Mr. Ray.[56] George Washington Jordan died from "congestion on the brain."[57] A Mr. Haslett died of an unknown cause, as did Mr. O'Ferrall, Charles Blake, Joseph Gratny, John Walsh and James Kibbee. George Long died as the result of drowning while crossing the Missouri River. Then, too, there were the deaths of Joseph Hempstead, attributed to cancer, and that of Augustus Coriell by explosion of a powder keg.

The larger the emigrant group, the more desirable it became to formalize regulations by which it would be directed and according to which decisions would be made, i.e., a system of government. To that end, the Dubuque Emigrating Association, which left Dubuque in the spring of 1850 and was then at Kanesville, saw fit to establish a compact by which its members would be governed during their overland journey to California. In it, the members addressed the issues of leadership, safety and mutual support for one another and elected a "surgeon." They also resolved that they were willing to pay fair and reasonable ferry charges but they were "not willing to submit to extortion." The document had forty-nine names of Dubuquers appended to it.[58]

The Dubuque Emigrating Company (a different faction with a similar name), traveling overland to California, began its journey in Dubuque in the spring of 1852, arriving at Kanesville in April. We know that the original group leaving Dubuque was composed of ten members of the Cooke family, twelve additional "young men" and four wagons. Additional travelers from Dubuque were added to the company at Kanesville, as well as some non-Dubuquers, to reach a total of seventy-two men, a size thought to be advantageous to the venture.

The list of names provided as signatories to the compact consists of men only. It is known, however, that others in the company did not sign this document. Counted among these were Lucy Cooke and her infant daughter, Sarah, as well as Lucy's mother-in-law, Sarah Cooke, and her children, Eva Anna, ten; Edward, seven; and Richard, five. Likewise, Susan Madeira (wife of George) and her thirteen-year-old son, George, were not counted.

It is interesting to note the differences between the compact agreed to by the Dubuque Emigrating Association and that of the Dubuque Emigrating Company. The latter gives much briefer attention to operational and security procedures while affirming religious and moral standards, an element totally lacking in the association compact. The company resolved to observe the Sabbath and to "avoid open immorality of every kind, such as profane swearing, gambling, and the use of intoxicating drinks as a beverage."[59]

Chapter 5

Except for the Gold, Not as Good as Iowa

Among the ideas advanced as rationale for the miners of the Upper Mississippi River Lead District abandoning the lead mines and heading out for the gold mines is that lead mining was very hard work. This arduous labor would be behind them when gold mining in California, so they presumed, since all one had to do was "pick up my share of the rocks."

With gold fever as the principal cause of the great migration to California, this supposed ease of gathering in the gold must have been a belief widely held. More likely, it was not so much a belief but, rather, a thing ardently desired. Once in the gold regions and actually mining for the gold, the folks from Dubuque provided dismaying testimony to the heavy lifting involved, often made heavier by the meager return.

After a journey of three to four months, and sometimes more, the forty-niners arrived in the gold region of California, tired, worn out, perhaps sickly and surely exhausted by the lengthy journey. Yet they had made it. The question then became, "What now?"

While many difficulties were encountered getting to California regardless of the route taken, hardships did not cease upon arrival in the gold regions. The California country was still relatively undeveloped, and many had to make do in crude, temporary shelters—tents, brush enclosures, the occasional boarding shanties and caves or simply a blanket or two under the sheltering pines. And, as the adventurers were soon to discover, gold was not just simply waiting on the surface to be scooped up by the bucketful.

One of the first tasks incumbent on the Dubuque forty-niners was to resupply themselves. They had encountered shocking prices on the trail and become somewhat accustomed to them. But upon arrival in California, they had to pay as much and even more for common items—when they could be had at all. Joseph Hempstead, writing in 1849, notes that Tom Childs, a friend of his from Dubuque then in California, sent him one hundred pounds of flour. The flour cost twelve dollars (two dollars in Dubuque), but the freight to get it to Hempstead at the mines was another fifty dollars. He also notes pork selling at fifteen dollars per hundred pounds (two dollars in Dubuque).

In another example, James A. Langton, writing from Sacramento to his brother in Dubuque, notes that he was seeing $100 for flour and $75 for pork; $15 per barrel for potatoes, and of "no account" at that. Finally, in the matter of eggs (twelve cents a dozen in Dubuque), J.P. Evans, writing on December 20, 1849, from Sacramento to his friend Platt Smith in Dubuque, notes that they were bringing $1 each egg, not each dozen.

Many more similar examples are found in the letters Dubuquers wrote to family or friends back home. The high price of living expenses was an unending source of consternation, certainly during the journey but likewise upon arrival in California. Letters from Dubuquers contained equal expressions of shock and dismay at the high costs, though one wonders if the men really had a close familiarity with food prices, this usually falling under the domestic purview of women. Men were forced to acquire this knowledge, there being relatively few women in California in these early years.

For the women who were in the gold regions, one concludes that they had just as much, and perhaps more, entrepreneurial spirit than the men. When it came to making a buck, they were right there—front and center. Once in California, the variety of their skills led them into occupations that were more secure and far less speculative than searching for gold, though some did engage in that enterprise.

One writer noted that if she had to do it again, she would borrow the money to get to California, as "it is the only country I ever was in where a woman received anything like a just compensation for work." Running a boardinghouse was not an unusual venture for women, but it would have taken some considerable resourcefulness to find a square of ground to occupy and perhaps a tent or two to start out. Other occupations included running a theater, barbering, taking photographs, nursing, teaching school, taking in laundry, sewing and making pies (at $1.25 to $1.50 each). One of the pie makers noted that she had made $18,000 worth of pies, netting some $6,000 clear for her labor.[60] Many of the gold prospectors would have been thrilled by that kind of success. There were also more sinister enterprises as well.

Dubuquer John Coffee, writing to E.F. Bissell in October 1849, noted that his wife and daughter were "engaged in washing and make from $12 to $18 per day" and that "a strong good girl can make a fortune in a year." Another Dubuquer, Edmund M. Allen, relates to his friend J.C. Weatherby that "a good steady female can make a fortune in a year at nursing or sewing if she keeps her health, but this is a rough place for any female to come to." It is clear that for many women, California was the land of opportunity.

Once the men actually began "picking up" their gold, they found it to be arduous work. For some, it was the most difficult work they had ever done. Joseph Hempstead, writing to Thomas McKnight in Dubuque, thought that a man could make money in the mines, but it was the "hardest work a man ever done" and one could stand it only a short time. William H. Merritt wrote in January 1850: "I returned from the mines the other day with seven hundred dollars which was obtained by the hardest labor man ever endured…and I could pray with the fervency of a saint that my punishment was at an end. Yet my mission is unfinished, and while that remains, I must remain to suffer."[61]

He reiterated the same message a month later in a letter to his brother in which he supplied examples of what hard work was: "Digging on the canal, quarrying rock under a broiling sun, or going into the cold water up to your waist before sunrise. Gold hunting is decidedly and emphatically the severest labor known to man." This is a theme repeated frequently by the Dubuque forty-niners.

In addition to the struggle to make ends meet, there were also health conditions to contend with. Cholera was a familiar companion to all those making the journey to California. Even having completed their journey, these travelers would not escape the attention of this disease upon their arrival in the gold regions.

Joseph Hempstead had a rather jaundiced view of the general situation regarding health once he arrived in California. He supplied a firsthand account that he and his fellow Dubuquer Tom Childs experienced. He believed California to be among the "most sickly" countries he had ever been in, with hundreds of forty-niners who would not live to return to the States—not knowing at the time that he would be counted among them. He asserted that half the people in the mines were sick and dying every day and would return home if only they had the means to do so. He and his friend Childs had been mining on the Yuba River for about three weeks when he was stricken by typhus fever and Childs with scurvy. He was unable to leave his tent for seven weeks and was not expected to survive. Childs was so bad that they didn't expect him to live either, so he was shipped off to San Francisco to get care, where he did eventually recover.

Cholera arrived in California not only from the overlanders but also from the Isthmus. In one representative example, it reached San Francisco in October 1850 on board the ship *California*, with twenty-two of its passengers suffering from the disease, eventuating in the death of fourteen of them. Other presumably healthy passengers of the *California* booked passage upriver to Sacramento, where one of them died from cholera upon arrival. The dreaded disease had arrived in Sacramento. In a period of four short weeks, about another one thousand had died. Fearing the worst, about 80 percent of the population took flight, spreading the illness even farther. It is reported that San Francisco lost 5 percent of its population, and as many as five thousand may have died in northern California.[62]

Scurvy was the other common health issue for the miners. William H. Merritt, writing from the Tuolumne River to his brother, pinpoints the problem: "A large portion of the miners here made their diet almost entirely of pork and sea bread, and this as a consequence produces scurvy, one of the most loathsome diseases on record."

It is understandable that there would be a lack of fresh produce while on the trail, but this deficiency was only slightly better once in California. The reader must recall that the Central Valley was, in 1849, mostly undeveloped land, with little of the great agricultural production that was to come in later years. Of course, the mountainous regions being mined were no better. The result was that after being on the trek west for some four months and then working in the mines, a vitamin C deficiency was found in many of the miners. Hempstead records a sobering description of this all-too-prevalent disease:

> *Quite a number of deaths among them—most all have the scurvy… Heacock who has the scurvy so bad he can't walk. Charles Lorimier has a touch of it also but he is able to be about. J. Wharton is here also with the scurvy so bad he can't move; lots of others in the same condition from our section of the country.*[63]

He summed up the general situation by declaring California one of the "hardest" countries he had ever been in. Many of the migrants had no sooner arrived than they sold their teams and headed back to the States without even going to the mines. Hempstead believed the reason for this was that they saw so much sickness throughout the country that it scared them.

For those who stayed and paid the high prices, worked the gold fields, endured the hard working conditions and attempted to persevere, moral support from the homefront was critical. Letters from family and friends were at the top of the list, but newspapers were also eagerly sought after and passed around from one to another.

Letters from loved ones in the States were frequently directed to San Francisco, where there were so many arriving that the few people working the mails there were simply overwhelmed. Hundreds of prospective recipients would form lines and wait for hours only to find that they had not received anything after all. Letters sent by the miners to family and friends back home cost one dollar postage per letter. The price seems high, but in the gold country, this amount was scarcely missed.

For many of the forty-niners, regardless of the amount of time they endured in their pursuit of gold, their thoughts frequently turned to home, family and friends. This was especially true for those who were barely making enough to survive. Adding to their declining mental state due to the realization that they were not going to get rich was the daunting thought of having to make the return journey back home.

Most had taken the overland route to California and knew it to be long, arduous and dangerous on a number of counts. Now they had to face reality and begin to plan for their return. It is not too surprising that, as Holliday notes, "Literally everyone returned to the States from San Francisco by ship. I know of no record of anyone traveling east on any of the overland trails…In 1850, departures recorded by the harbor master at San Francisco totaled 26,593."[64]

Ships aplenty had by that time entered the trade where they had been sorely lacking in 1849, though seaworthiness was sometimes questionable. Routes across the Isthmus of Panama had improved, the building of a rail line had commenced between Panama City and Colon (completed in 1855) and the cost for passage had become somewhat more reasonable. Taking this route back east was much preferable to the overland journey and could reliably be completed in five to six weeks, compared to the four months on the overland trail.

The conditions making this quicker journey possible had developed within a relatively few months after the inception of the great migration to California, and this was just one more way for an entrepreneur to cash in on the gold rush. In the end, many who had gambled their savings or borrowed money to get rich quickly had to swallow their pride and return with little or nothing gained.

Chapter 6

By Sea to the Gold Fields

After the decision had been made to join the California gold rush, there were yet more choices to be made. *How much will it cost to get there? Where will I get that kind of money? When will I leave? With whom will I travel? What route will I take?* These were only a few among the many questions to be faced and answered.

Word-of-mouth reports of the gold discovery in California likely arrived in Dubuque in August 1848 due to its commerce with St. Louis, which had the news at that time. There was, however, considerable speculation about the truthfulness of the reports. News reports appeared in the Dubuque newspaper in October but were downplayed by the editors. The confirming news that the discovery of gold in California was to be believed arrived in Dubuque in December 1848.

The winter season was then upon Iowa, and overland travel to California was out of the question until after the spring thaws, the arrival of warmer weather and the growth accompanying the new season. Some, however, had the gold fever to such an extent that waiting was not an option, for the longer one delayed, the more likely it became that someone else would be picking up *his* gold. And so it was that they began looking for other ways to get to California sooner.

This led to the selection of a course of travel that would take them south to St. Louis and on down to New Orleans. From there, transportation could be found across the Gulf of Mexico to Chagres on the Caribbean coast of the Isthmus of Panama. A relatively short but quite unpleasant transit across the isthmus brought them to Panama City on the Pacific Coast. The plan was to then travel via steamer up to San Francisco. As we will see, the plan ran head-on into reality, much to the dismay of the Dubuque forty-niners.

William H. Merritt, who was then the editor/proprietor of the Dubuque *Miners' Express* newspaper, along with Edward Mobley, S.M. Hammonds and J. McCoy, made the decision to leave as soon as possible, which they did on January 1, 1849. In Merritt's letter, dated March 9, we find also on the isthmus from Dubuque were Augustus, John and Edwin Coriell, J. Douglas, J. Clark, J. Hempstead and a Mr. Marvin. These folks were determined, and had the wherewithal, to leave without waiting for spring to arrive and with it, the availability of the overland trail to California.

Merritt was born in New York City in 1820 and was educated at the Genesee Wesleyan University in Lima, New York. For a number of years, he engaged in a variety of mercantile ventures. In 1847, he took over the Dubuque *Miners' Express.* In addition to his newspaper work, Merritt was a prominent Dubuque citizen and lawyer. He had no experience in lead mining and likely as little in gold mining. But at the age of thirty, though married, he succumbed to the gold fever.

After returning from California in 1851, he resumed his association with the *Miners' Express* and, in 1852, was appointed surveyor of the Port of Dubuque. He had again left Dubuque by 1855. In 1861, he became the Democratic nominee for governor of Iowa, for which post he was defeated by Samuel J. Kirkwood. He served as lieutenant colonel in the Civil War and fought in the Battle of Wilson's Creek. In 1880, he became mayor of Des Moines. He was married and the father of seven children.

The sights and sounds, the troubles and tribulations, the despair and jubilation of this forty-niner come to life most vividly as told in the first-person narrative of his letters. With these, we gain an understanding of the anxieties and sense of enthusiasm experienced by this Dubuque forty-niner.

William H. Merritt to his brother George, February 4, 1849, on the ship *Wm. H. Hazzard*, 450 miles from New Orleans:

> *My Dear Brother,*
> *Sufficient time has elapsed, no doubt, for you to expect another letter from me. And perhaps you may feel disposed to accuse me of a little neglect that I did not write you from New Orleans. I was there but two days, every moment of which time I was engaged with Mr. Titus in making our purchases, as we were selected for that duty. While at Mr. Jones' grocery establishment, purchasing our groceries, I did steal a moment to write to* ___ [sic].

I will now give you a brief history of my journey to this point, its advantages and disadvantages. Our expense in getting to St. Louis, was $22 each. We remained in St. Louis one week, on account of cholera in New Orleans, at an expense of $9 each. I find that our southern brethren understand one principle of trade, that is, how to skin the stranger.

...Our trip down the [Mississippi] *river was not of the most agreeable character. We took passage on the steamer* Alice, *for New Orleans, at a cost of twelve dollars, cabin passage. This, I believe, is the cheapest line of communication in the country. Our route down the river was a pleasant one. The land on either side the river is richly stored with heavy timber. When we left St. Louis, the ice was running thick in the river. This we found a great encumbrance to our progress. In place of leaving it behind us, as we expected, the farther south we went, the more obstinate became its resistance, until we arrived within twelve miles of that scriptural city, Cairo, when we were suddenly and completely shut out from the world below—at least by steam.*

Previous to leaving for Cairo, we settled up with our gentlemanly clerk, and received eight dollars of our passage money back, in conjunction with a thousand wishes for our health, happiness, and success in California.

We placed our trunks on board of a cart, hauled by two yoke of oxen, at a cost of 50 cents for each trunk, and commenced our march on foot for the city. We arrived there a short time after meridian, and some two hours in advance of our stout-hearted ox driver. The city forms a right angle, and is built upon a low piece of ground; the larger portion inundated the year round. The buildings are of wood, rough in appearance and fast going to decay. They are protected from daily flood and destruction, by heavy banks thrown up on the Ohio and Mississippi.

The people are indolent, dissipated, and sickly. Notwithstanding the saintly name given to the place, its inhabitants seem to have forgotten the sanctity which surrounds its early history, in their habits of business and intercourse with the world. I have never visited a place, or become associated with a people, more repugnant to my feelings than Cairo and its low-flung inhabitants.

Withal, they are a somewhat speculative people, and as an evidence of this fact, I will relate a little circumstance which occurred to mar the equanimity and Christian forbearance of our little company.

You will recollect that I told you in a former part of this letter, that we negotiated the hauling of our trunks at 50 cents each:—so we did. But when our avaricious jockey of the saintly city made his appearance, and manifested his readiness to receive charges and deliver goods, the freight had advanced from fifty cents, to one dollar on each trunk, and a stout refusal to deliver the trunks till the full price claimed should be paid.

At first we thought it advisable to reason with the fellow, but finding that this only inspired him with fresh courage and obstinacy, accompanied with repeated heavy claps of thunder, in the shape of oaths and imprecations, we determined to show this rude genius and his companions what virtue there might be in physical force.

*Every man, quick as thought, mounted the cart, and each one laying violent hands upon his own trunk, tore it from its covert. In the general melee, Mr. Ox Driver got tumbled head-over-heels into the mud, by that accident (*if *accident) or by whose rude hand, I know not; and if I did, it is very questionable whether I would name the culprit. It required but a moment for each man to fling out his half dollar, and order his trunk aboard of the* Constitution, *where our fare had been previously engaged for New Orleans at an expense of eight dollars, making, in steamboat and land carriage from St. Louis to New Orleans, even money, $12.50.*

...We were six days performing the trip from Cairo to New Orleans, including four or five quite lengthy delays. From Cairo to Memphis, I was not particularly struck with the beauty of the country we passed through, though it may be a poor judgment, based upon a careless observation of that part of the country immediately bordering upon the river, which you know, was all that I had the advantage of seeing.

...We arrived in New Orleans on the 25th January; and being compelled to leave on the 27th, you must be aware that I had but little time to devote to acquiring a knowledge of the place and people. However, as I passed from house to house, and street to street, in pursuit of various articles constituting our outfit, I was not a careless spectator of the various and peculiar scenery which presented itself to the stranger's view.

It is a place of great commercial importance—commanding, as it does, an extensive foreign and inland trade. Its coast is lined with shipping for miles, and presents a most magnificent spectacle from both city and river. Such bustle and confusion as the levy presents, with its ten thousand clerks, boatmen, draymen, and porters is seldom witnessed in the larger commercial marts of the North-East.

The ground upon which the city is built is low, wet, and marshy, the streets generally narrow, the buildings of good quality, and of English architecture. Great expense and pains are bestowed upon drinking houses, and they are found in great abundance in all parts of the city. Many of these rooms are spacious, richly ornamented with French and Italian paintings, and plentifully stored with the finest of foreign wines and liquors.

The people are a heterogeneous mass made up of subjects from every clime and country.

...In my heart I could find but one objection to them, and this perhaps I might excuse, were I not committed as one of its opponents. They drink too much strong "tea." *It is used indiscriminately without regard to class, condition, or age. The youth of twelve quaffs as full a glass, and supports as fine a cigar, as the man of wealth and fashion.*

[concluded next week][65]

William H. Merritt to his brother George, February 4, 1849, on the ship *Wm. H. Hazzard*, 450 miles from New Orleans:

My dear brother,

...

To our little bark then, I will conduct you.

She is a two-masted fore and aft schooner of 212 tons burden with a full crew of raw, two-and-a-half percent Yankees. We have 74 passengers on board, 23 in the cabin and 51 in steerage, all bound for the gold mines of California. We left Orleans on the evening of the 27th January in company with the steam tow-boat General Taylor *and arrived at the Gulf the next morning about 8 o'clock, a distance of 150 miles. This was fast running, too fast for any* General *except* Taylor, *to accomplish.*

...As the steamer cast off our cable, up came a smart boat moved by four men at the oars and one at the helm. As the little craft came alongside, the man at the helm, who was a short strait and remarkably pompous little seaman of about fifty, hailed us by inquiring if we wanted a pilot. The captain answered in the affirmative, when the little fellow mounted the ship's side, planted himself upon the hurricane deck and in a gruff tone ordered the men to haul in the cable.

This order not being obeyed as readily as he desired, he sang out again, haul in that cable. Some of the smaller ropes having got tangled with the cable, it required some little time and exertion to prepare for shipping it. This excited the pilot beyond his capacities to endure and just as our willing crew commenced shipping the said cable, he broke forth in angry tones "G-d D--n your lubberly pates, ship that cable or I'll run your old flat-boat spang on to the reef."

This elicited from the sailors some little murmuring, but they finally obeyed the wink of their captain by redoubling their exertions, shipped the cable, hoisted all sail, and with the little pilot at the helm, were soon safely beyond the reef, when the captain resumed command of his craft and steered a south-easterly course with the wind dead ahead.

For seven days were we beating about in the Gulf, making perhaps 40 or 50 miles in twenty-four hours. On the evening of the 7th day however, the wind shifted around to the north-west, and our little craft bounded through the water at the rate of 11 knots an hour, which brought us in sight of Cape St. Antonio the next morning about 10 o'clock. We doubled the Cape about 3 o'clock p.m. and sailed along the south coast of Cuba, some two miles from the land.

The greenness of the grass and leaves upon the trees and the fragrance which they omitted were realizations which a man from the North, and a stranger to the sea, eight days from land, could not but experience the most pleasurable sensations. We had a fine summer breeze for running down the coast. I remained upon the deck until a late hour in the evening and even then found it difficult to take my eyes from a scene which commanded every poetic impulse.

I slept but little that night, owing either to the heat of the climate, which I find lessens the quantity of sleep required, or a desire to pursue my observations which were necessarily broken off the preceding evening. The earliest dawn found me upon deck and in conversation with the pilot.

Over our larboard [port] *bow and some twenty miles down the coast, he pointed out the Isle of Pines, with its three sugar loaf mountains, St. Jose, Canada, and Dagilla. This Isle of Pines you must have remembered for its piratical notoriety. No place in the world stands more prominent in naval and nautical history as a retreat for pirates than the Island of Cuba, and no place upon the Island than that of Pines.*

We passed down the coast keeping in sight of land all day. After the sun disappeared a fresh breeze came from the land and we tacked ship and bore out from sea. The breeze freshened as the night advanced, until 10 o'clock, when a large black cloud in the north-west betokened an approaching storm, and the ship's crew were all summoned to the deck. Presently the wind howled, large drops of rain commenced falling, a red glare of fire streamed in quick succession through the troubled firmament, and the white caps rolled mountains high threatening our little bark with speedy vengeance from heaven's artillery.

For two hours did the storm rage with unabated fury, towing us about from place to place and creating among a portion of our California boys a terror and excitement not expected from those who had voluntarily undertaken an expedition to that distant clime. By and by the sky broke away, the wind subsided, and after a time the sea rolled down. No one interposed an objection, although it left us in almost dead calm.

The 10th a somewhat novel occurrence took place among us. Some one hollowed out land over the bow. A rush was made for the deck. A division of sentiment was at once manifested as to the truth of the report. The sides were about equally divided. From a pleasant difference of opinion soon arose a lively excitement and an offer to bet fifty dollars was offered and taken, twenty-five was also offered and taken—numerous bets from one to ten followed in rapid succession.

The excitement for four or five hours was intense. Each side manifested equal certainty of success. The appearances fully justified this state of feelings, for nothing ever appeared better calculated to deceive. To many it appeared like a huge dense mass of black clouds piled upon each other thousands of feet high. To others equally sanguine it appeared like mountains rising above each other until they appeared to kiss the very heavens. For five long and tedious hours were we in doubts to this singular phenomenon.

At length it became apparent that we were approaching land. How quick did the countenances of our little company change. The change was sudden and complete. Where doubt had marked its victim, success now lent a playful smile, and where joy a moment before had gleaned in hopeful security, disappointment and despair now found its way.

We had approached within ten miles of the southern coast of Cuba and could plainly discern the city of Trinidad. We ran to within two miles of the shore and could measure with reasonable certainty the dimension of the city, and the orange and cocoa nut groves. In this part of the island there are also large groves of cedar and mahogany.

Map of Gulf of Mexico and Caribbean showing waypoints to the California gold fields. *Courtesy of the author.*

...Trinidad is a place of about six thousand inhabitants, is situated upon an elevated piece of ground, commands an extensive view of the sea, and is surrounded by mountains of from six to eight thousand feet above the sea.

I had almost forgotten to mention that Trinidad is also noted for its fine crops of sugar and coffee. I shall now indulge a short respite from my labors. It is to give our little bark a chance to reach Jamaica where we shall anchor for a day or two where I will resume my task. Should the wind favor our design of stopping at Jamaica, it will be acceptable to me on more accounts than one. First, because I am tired of the sea, second, because I wish to procure some fruit, and last, but not least, because it will afford me an opportunity of mailing this letter to you and one to my dear _____ [sic]. *So farewell until I reach Jamaica.*

We reached the west point of Jamaica on the 13th. We drifted around the western coast of that island until the evening of 15th, when a brisk land breeze sprang up and drove us off in the direction of Chagres.[66] *While in the vicinity of the island we had no wind favorable to making a landing, so that I was disappointed in paying court to the sooty lords of that rich and fertile island. I saw nothing which particularly attracted my admiration, except the range of high mountains upon the coast, so wisely calculated to protect the island from inundation.*

I am told that there has been a sad falling off in productive industry upon the island since the abolishing of slavery. The Negroes have become reckless, indolent, dissipates. So much for that British philanthropy, which frees its foreign blacks and enslaves the domestic whites. Their Negroes of Jamaica are converted into landlords and stock jobbers, while their own brothers at home are never permitted to escape the eternal darkness of their mining dungeons.

This is modern philanthropy among the English. Their political avarice has led them into an error degrading and foolish. Had they attempted to free the women and children born to slavery upon the British island, in place of the Negroes of Jamaica, that public gratitude which they have failed to achieve would have been cheerfully and freely awarded by a liberal public. But they have sought to steal a reputation for justice by pursuing the very course which renders it unattainable. They have signally failed in securing this character in every other place except "ome."

But let us drop Jamaica, its Negroes and former masters. We arrived in Chagres on the 20th all well except four who had been sick from the time we left the dock at Orleans. Their complaint; sea sickness. Not a dangerous disease, although two of the number exhibited all the grautic fear of men about to "give up the ghost." They are bad stock for the California market.

Well, what shall I say of Chagres? If I should pronounce it a fine healthy place inhabited by men with heads above their shoulders you would scarcely give me credit for truth, and yet I must tell you in all sincerity, that very much said in disparagement of Chagres and its people is untrue. It is in latitude 9.27 which to people born and reared in a northern climate must be unhealthy. But the idea as stated by many writers that men from the United States cannot remain there twenty-four hours without being surrounded by crocodiles, serpents and lizards, or attacked by some pestilential disease which no medicine will cure, is a foolish humbug.

True it is low ground upon which the town is built, generally muddy on account of the great quantity of rain that falls, and inhabited by black skins of an inferior quality, but yet with proper care, no one need to fear to stay there for a few weeks during the winter months. People from a cold climate would be hardly justified to making a halt there during the months of July, August, and September. The population of this Negro village is about 400. My time will not permit of my writing more at this time. I will resume my epistolary labors at Panama [City], *where I shall have more time and be in possession of more information upon which to write.*

Yours truly,
W.H. Merritt[67]

William H. Merritt to an unnamed addressee, March 9, 1849, from Republic of New Granada,[68] Panama City:

My Dear _____,
My last letters were mailed at Chagres to yourself and my brother George. In one of those letters, I promised to write again at Panama [City].[69] *I have now seated myself to fulfill that duty. Let me first inform you that my health was never better, and that I was highly gratified at meeting Mr. Peckrem yesterday and learning from him that you were well when he left Dubuque on the 15th January, and that my dear sister H. had so far recovered as to be able to walk out. I must acknowledge that I had many fears when I left home as to her recovery. I trust she is now out of danger.*

Oh! Could I know from day to day that my family and friends were in enjoyment of good health I could forget the pleasure of their society without a murmur, but when I reflect that I have had no intelligence from you for two long and dreary months, and that I am to remain unadvised of your situation for probably two months longer, and possibly four, I do assure you that a gloom passes over my mind which is anything but pleasant.

After leaving Chagres, I was three days getting to Gorgona,[70] *which is situated on the Chagres River about 45 miles from the first named place. I was agreeably surprised in the appearance and character of this river. From the representations in the American prints I had expected to find a low, dirty stream of water filled with reptiles of every description, but on passing up the river in steamboat and canoe I found that my mind had been abused upon this subject.*

A more beautiful stream of water I never beheld. The stream is very rapid, clear and pure. This water is of a fine flavor and would be exceedingly palatable if it was cold, but the uniform hot weather renders it, like everything else, very warm—the temperature of the atmosphere being from 85 to 95 degrees of heat. The scenery along the river is delightful. As you pass up the river your observation is attracted by a high range of mountain lands on either side, covered with a luxuriant growth of mahogany, cocoanut, bananas, plantain, wild sugar cane, cotton, rice &c., &c.

March 15th. Parrots and paroquets are playing among the trees as plentifully as black birds in our country; monkeys are yelling in every direction and now and then you espy an alligator stretched upon the bank, sunning himself. I saw two on my journey, I should judge to be 10 or 12 feet long.

...I remained about a week at Gorgona—a small village about the size of Chagres inhabited entirely by natives. During that week I made a trip to Panama [City] *at the request of my company to secure, if possible, our passage to California and ascertain whether it would be advisable to attempt to get our provisions over the land route. This journey of 22 miles was made on foot in seven hours over as rough, rocky, and mountainous a route as it is possible to imagine.*

I arrived in Panama [City] *about one o'clock, p.m., and although completely exhausted, I assure you it was refreshing to find 8 or 10 of my old Dubuque friends,* viz., *August, John and Edwin Coriell, J. Douglas, J. Clark, McCoy, J. Hempstead, &c. I remained in Panama* [City] *one day, secured our passage at a bargain ($200 each, in sail vessel), became convinced that we had better dispose of our provisions for whatever I could get for them at Gorgona rather than pay twice their original cost for transit across the land, and returned to report what I had done. After hearing my report it was resolved that myself and Mr. Marvin should remain at Gorgona until we could sell our provisions, and the balance of the company should start for Panama* [City].

We remained four days, during which time we disposed of our provisions and left for Panama [City]. *When we arrived at the city we all got together settled up our accounts and after paying the expense of getting our provisions to Gorgona, each member of the company received $5.85. So you see this was rather a poor speculation.*

I have now been in this primitive city 8 days on an expense of from 50 cents to $1 per day. Everything in the provision line is very high and it is only by hiring a room in company with ten others and living upon the very plainest diet, that I manage to get along so cheap. There are many here who spend not less than $20 per week for board. The public houses charge from $2 to $3 a day independent of wines. The ship in which we have secured passage is expected on the 15th. She will remain in port 2 or 3 days, and then we shall be off for San Francisco. It is expected that we will be from 40 to 60 days in running up.

The steamer Oregon *left here this morning with 238 passengers. The* California *is daily expected. There are here more than she can possibly get on board who secured their tickets in New York in December 1848 and at Chagres, and on the way between here and there it is estimated that there are twice as many as the* Oregon *can possibly take on her return. There are from 800 to 1,000 Americans in this city, and it is thought from ten to twelve hundred at Chagres and on their way to this city.*

Two Americans have died since my arrival in this city from fever produced by exposure to the sun and a too plentiful indulgence in the tropical fruits which are considered, even by the natives, injurious. There are now some 12 or 14 sick. With ordinary prudence those attacked by the fever recover. I am only surprised that many more are not sick. I am assured that the same exposure in their own climate would prove far more fatal to them.

For one, I could not desire a more healthy climate. But you must bear in mind that this is the dry and the healthy season. From the latter part of November to the middle of April there is but little rain, and therefore but little sickness; but the balance of the year is one almost uninterrupted season of rain, and therefore a great deal of mortality. A French doctor here tells me that not more than one out of four Americans survives, who ventures to remain upon the Isthmus during the entire season.

Panama [City] *is a walled city and in its more prosperous day and generation was no doubt one of the most impregnable fortresses of the Spaniards. I send you a copy of the* Panama Star, *an American paper published here which contains a brief description of the city, and will therefore relieve me from a detail of its character.*

What can I say of the inhabitants of this little neck of land? Certainly nothing in their favor. They are a mixture of Spanish, Indian, and Negro—the latter predominating to a very considerable extent. They are a proud, ignorant, dissipated, treacherous, and dying race, and I blush to admit, that little or no effort is made by the civil or ecclesiastical authorities of the country to mitigate this evil. The most hate-faced frauds are practiced upon the Americans, and the public authorities neither disapprove nor punish.

This species of outrage has become so common of late, that the Americans have finally been thrown upon the dangerous alternative of assuming the functions of judge, jury, and executioner. They have taken the law into their own hands, and when an imposition is sought to be practiced upon them, they chastise with becoming severity. It has had a most beneficial effect upon the American interest, for it has taught the natives that there is a superior race of men to themselves, and that while they respect the laws, civil and social, of a foreign nation, they will not allow their property to be wrongfully taken from them, and their persons insulted and out raged with impunity.

Of late, they are as timid of the Americans as a child is wont to be of a cold, misanthropic parent. In fact, they are absolutely afraid of them and how should it be otherwise? They behold a thousand or fifteen hundred men and boys, literally loaded down with guns, pistols, and bowie-knives. Really, it is farcical to witness the enormous loads that old men of three-score, and boys of tender years pack about with them.

Wm. H. Merritt[71]

William H. Merritt to Dr. H. Holt, March 29, 1849, from Panama City:

Friend Holt,

When I wrote the within letter, I expected it to have gone by the steamer Northerner *which sailed from Chagres on the 24th inst.; but the passengers for that boat left this city about three hours sooner than was expected and therefore disappointed me. Another opportunity however has occurred, and I improve it, to add a few lines in commemoration of my letter which by the time you receive it, will be old enough according to usage, to have died, been buried, and epitaphed.*

...Had we all taken more time for reflection, and instead of rushing into the current of universal excitement which flooded the country at that time, provided means for an early start by land, we should not have been swept "Crusoe" like upon this barren rock to while away weeks and months in dreamin' about the gold mines of California, without a possibility of reaching them before "war, pestilence, and famine" overtake us.

Did I not dread the idea of turning back, I should be almost tempted to go [back] *now that we have lost our passage upon the* Orion. *It was an old paternal injunction never to look back after I had once taken hold of the plow. Up to this time, it has been faithfully observed, and I think you may set it down as a fixed fact, that I shall not disregard it, even here, penned up as I am with cannibals and monkeys, lizards and crocodiles, white Negroes and black, until I have exhausted every possible resource for going ahead.*

Americans are flocking in here by hundreds daily, to spend weeks and perhaps months in this restless and dismal hole, where the habits of the people are as little adapted to elevate and amuse the mind as the climate is to invigorate the body. I have somewhat seriously felt the debilitating effects of the latter, and I fear, if I am forced to quarter here much longer, I like many of my frolicksome countrymen, shall not escape that of the former. There is considerable sickness among both natives and Americans.

Many of our friends too are out of money. This is a most lamentable state of things, but I know of no people whose capacities are better adapted to overcome this difficulty than the "Universal Yankee Nation." Their genius is certain to shine in this sphere of enterprise—and where a savage can live, they are certain to thrive.

I have not time to write more. Excuse mistakes, and believe me your obedient servant and faithful friend.

Wm. H. Merritt[72]

William H. Merritt to Dr. H. Holt, April 5, 1849, from Panama City:

Dr. Holt,

One disappointment seems to follow another in rapid succession. Mr. P. [Perrin] *has changed his mind and concluded to remain here and take ship.*

The Americans continue to flock in by scores. Many are leaving, too, for the United States and South America. The Northerner, *on her return had over 100, the English steamer which left here a few days since for Valparaiso and Callao had 196 and could have had as many more if she would have taken them. The* Crescent City *took back 100 of her passengers who never ventured a step into the interior.*

Through the instrumentality of the American Consul, myself and Dubuque friends (Mr. P. included) have secured a passage upon the whale ship Niantic *which is expected to arrive here in two or three days. It was a piece of good fortune that Mr. P. got on the boat, for passengers were taken according to the date of their arrival on the Isthmus, and none who arrived since the 20th February were received except him.*

The English brig Two Friends *arrived in port yesterday—being the first sail of any description except an English man-of-war that has darkened this bay in four weeks. She was snatched at $250 steerage and $350 cabin within the twinkling of an eye. She proposes to sail for San Francisco in 15 days.*

She takes 150 passengers—too many by 50. This crowding of boats with passengers beyond their capacity to render safe and comfortable, and contrary to express prohibition by law, should be carefully looked to by the agents of the government in California. The law grants vessels 1 passenger to each 2½ tons of register. Not a vessel has left this port within 2 months (unless it be the steamer Oregon*) but what has violated this law. The* Equator, *of 260 tons, had 138 passengers; the English bark* Callooney *of 240 tons, has 150 passengers; the* Felix *of 40 tons, had 50 passengers; the* Constellation *of 20 tons had 30 passengers, and the English brig* Two Friends, *now preparing to sail for San Francisco, takes 150 passengers, with only 206 tons register.*

...There is one other subject which seems to require some notice. There are a few Americans here who are endeavoring to enrich themselves by speculating off of their countrymen unexpectedly detained upon this Isthmus. Generally they are men of no character at home, and when in a foreign country, and particularly a country where laws and language can take no note of their conduct, the restraint which law and public sentiment impose at home is laid aside, and they rush into the vices of trade and catch their prey with as much apparent eagerness and self approval as the man who is about to extricate some unfortunate fellow creature from a violent death.

To this class of "wandering Jews" we find now and then a convert who combines character with capital and becomes the Shylock for a season only. James L. Treaner, the matchless descriptive writer whose glowing portraitures of the great battles in Mexico obtained such unbounded popularity and renown in the New Orleans Delta *over the signature of "Mustang," I am sorry to say is one of this number in company with two other gentlemen.*

He has chartered the English brig Two Friends *to load with passengers and go to San Francisco and has fixed the passage at most exorbitant prices—$250 in steerage and $350 in cabin. A price which but comparatively few can pay, and that few by all usages of justice and humanity should be the last to receive the favor because they are the shortest time upon the Isthmus.*

Seven-eighths of the passengers upon this craft have not been on the Isthmus two weeks, but being fresh from the States they can afford to pay large prices, and by that means exclude men who have been here from six to ten weeks, and become so reduced in pocket as not to be able to reach the high mark at which they put their passages.

...Among the distinguished personages here is the accomplished Mr. Fremont, Col. Weller of Ohio, and Major Sewal—the first named bound for San Francisco and the last for San Diego.

Yankee enterprise is beginning to be seen and felt in this city. Since the Americans arrived here a new impulse is given to trade and commerce. Some have gone to Valparaiso and Callao to charter vessels and bring them here for the San Francisco trade—while others are engaged in mercantile, forwarding, commission, money exchange, and other similar transactions. Go into the principal street of this city and you will find all kinds of trades and professions in full blast. No less than three auction shops (a business not known here a month ago) are open and hammering away.

A month ago the water to supply the city was brought in on the backs of mules, but now a large cart, drawn by a yoke of oxen and commanded by a regular "down Easter" is doing a large business in the water line. A thorough reformation in the financial and mercantile transactions of the city has taken place—French and Spanish coin, which a few days since commanded a premium from 15 to 25 per cent, is now only worth its face; and American coin, which was greatly depreciated is now worth a premium.

Dry goods, groceries and provisions are rapidly coming down to a legitimate level, and everything here is fast assuming an American aspect. Wherever the American goes he is certain to leave the impress of his genius for enterprise and healthful competition.

The steamer California *has been expected for ten days but has not yet arrived. Many think that she is either unable to procure fuel at San Francisco or that her crew have left her, and she will not be here for months. The steamer* Panama, *on her way from New York, is expected here on the 17th inst. There are more than both steamers could take, if here, who bought tickets in New York.*

...Allow me in concluding this letter to correct one error which is very general among the Americans here, and I presume is the same in the States, to wit: that the rainy season is the sickly season in this country. It is a mistake. I was conversing last evening with Mr. Farand, a very wealthy and intelligent merchant of this city, and formerly American Consul for this

place, and he informs me that it is a very erroneous notion people contract with regard to the cause of sickness in this climate.

He tells me, contrary to all information I have received upon the subject, that the wet season is the season for health. That there is comparatively no sickness, and that those who are ill when the wet season sets in generally recover. He says the air is both cooled and purified, which you know is the fact, and the large quantities of filth, which is thrown into the streets and not taken up by the buzzards, is washed down the gutters into the sea. This will certainly be surprising to most people in the States, and doubtless gratifying to those who propose visiting California by the Isthmus route.

...

Wm. H. Merritt[73]

Merritt and his friends were still in Panama City when he wrote again on April 26, 1849. They were anxious to get off the isthmus in order to stop the drain on their financial resources and, of course, to get to California as soon as possible—before all the gold was gone. They had, by this time, been traveling almost four months already and were still nowhere near the gold fields. Having arrived at Chagres in early March, they were now approaching a two-month stay in Panama. They were quite frustrated about their inability to get moving and anxious about their future prospects in California disappearing before they even had a chance of achieving their golden dream.

William H. Merritt to Dr. Holt, April 29, 1849, from Panama City:

Dr. Holt,

Some matters of public importance having come to my knowledge within the last two days, I have thought it might be a subject of great interest to your many readers to become acquainted with them, and I will therefore proceed first, to give the substance of a conversation I had with Col. Weller, one of the Commissioners for settling the boundary between Mexico and our government.

Having occasion to call on Col. Weller yesterday to transact some business, I took the liberty of making some inquiries respecting the probable effect that a failure on the part of the Commissioners for the United States to meet the Commissioners for the government of Mexico at the time stipulated in the treaty would have upon the affairs of the two governments. His opinion was that it would produce another and a more bloody war than that just terminated.

His reasoning is as follows—that Mexico will be very glad of an excuse for repudiating the treaty now that the precious metals have been discovered in such abundance in California—that Santa Anna is feeding the flames of another revolution in Mexico, and that he is using the California gold mania as a means of bringing about that end.

The day fixed upon in the treaty for the meeting of the Commissioners is the 30th of May, and unless the steamer Panama *arrives by the first day of May, which is only two days distant, the Colonel informed me he should dispatch a messenger for the United States informing the President that he has failed, but that he would proceed to San Diego as soon as he could reach there, and use all his powers of entreaty to induce the Commissioners to overlook a delay entirely unavoidable. The Colonel feels very acutely the peculiar position he occupies, although no cause of complaint can lie against him, for the reason that he has acted in strict conformity with the instructions of his government.*

He was instructed to proceed with his Commission to Panama [City], *and await the arrival of the* Massachusetts, *which was subject to an order sent to Valparaiso directing her officers to go to Panama* [City] *and take on the Colonel and his party. The order did not reach Valparaiso until after the* Massachusetts, *had left that port for San Diego, so that the Colonel is left here without scarcely a hope of reaching San Diego in time to save the treaty.*

These are the facts, thrown together in a great hurry, and in a very crude manner; fix them up in "ship-shape" and give our friends the benefit of them.

I send you a copy of the Panama Star, *containing a brief account of a quarrel which took place at a* fandango *some evenings since, and which has produced a very unfortunate state of feeling here. The Spaniards were very much incensed at the conduct of some rowdy Americans there, and they have threatened all Americans with vengeance. This has put the Americans upon their guard, and they go armed to the teeth, night and day. I look for trouble to spring from this affair.*

The cholera is raging in Chagres to an alarming extent. I understood last evening late, that three cases had occurred at Gorgona. If this last report be true, it will be in Panama [City] *in less than 48 hours.*

About one hundred Americans are sick here with the fever; six died and were buried yesterday, and I understand five more have died today. The Niantic *sails tomorrow evening. We all go aboard tomorrow morning. We shall not be loath to bid Panama farewell.*

Wm. H. Merritt[74]

William H. Merritt to George Wallace Jones, July 9, 1849, from San Francisco, upper California:

Friend Jones:

I cannot think of leaving this city without dropping you a few lines although it is the second or third letter I have written to you without anything in reply. I wrote you before I left Dubuque and when at St. Louis, but I suppose the public interest is to be served, before private and humble individuals like myself, so be it—I am too good a democrat to complain.

I arrived in the Isthmus the 10th [actually the 20th] *day of February, left there the 1st of May, and after a passage of sixty-five days on the whale ship* Niantic, *landed in this mushroom city on the 6th of July. I have had a long and certainly a very tedious journey—expensive and aggravating. But I am rejoiced to inform you that reports from this gold region cannot be exaggerated. This whole country is a perfect mine of gold, if the statements of fair spoken men are to be relied upon.*

One thing is certain, that every man who has been here any length of time is loaded down with money—the observation that "gold is the cheapest article in town" has become perfectly stereotyped. Men have so much money with so few opportunities for spending it legitimately that they throw it away upon gambling tables, in grog shops, and useless merchandise that they purposely destroy in fits of drunkenness.

One man will buy a basket of champagne, go into the street and deliberately break the bottles to pieces; another will buy a dozen boxes of fine crackers and burn them. In fact the facilities for spending money are entirely too contracted for their means and their inflated minds. Goods are remarkably cheap and labor enormously high. The City Hotel rents for $144,000 per year, and other buildings in proportion. This I know to be true.

...Foreigners are thronging this country by thousands, and load themselves with gold for a foreign market. Unless the government speedily take some steps to correct this evil, war to the knife, and from the knife to the hilt, will be the result. Great uneasiness and jealousy begins to prevail already among the Americans, but they are too weak to move with any degree of success. The foreigners are at least ten to one. The English, our national enemies, are among the first to curse our country and steal our treasures. If an affinity has heretofore existed between them and the animal with horns, it now exists to a still greater degree between them and the brussels gentry; in this country

they are as proverbial for hoggishness as in their own for beef eating and ale tippling. I confidently trust that some measures will be speedily adopted by our government to prevent the exportation of gold to Europe.

Let me inquire what right the Post Master at this place has to charge ten cents for advertising a letter, or to shave gold by only allowing $4.75 for English sovereigns. This is his custom and is one which no law contemplates to my knowledge.

About 150 vessels are in this port, destitute of hands, and more arriving every day.

I leave for the mines day after tomorrow.

Haste and bluntness. Write soon.

Your devoted friend,
Wm. H. Merritt[75]

William H. Merritt to Mrs. Merritt, October 28, 1849, from the Tuolumne River, California:

I am in good health and hard at work living on pancakes and dried apples. Is not this a field for enjoyment! Surely this is the land for a Stranger*! Among rocks, mountains and grizzlies and men made hideous with beard and rags, entirely beyond the light and smile of a female countenance—who could forego the* pleasures *of a season in California! Ah, I fear yon gentle mistresses of the parlor will cry sour grapes.*

I have written you many letters since I arrived but know not whether you have received them. Not a line have I seen, nor word have I heard from you since I left San Francisco in July. It is an awful suspense. I pray God the pestilence, which has made such havoc in my poor country the past year, has left you and my friends unharmed.

I have just met Mr. Hadley from Dubuque. He crossed the plains. He has just communicated to me the melancholy intelligence of the death of Mr. Coriell. I was never more shocked. I had messed with him at Panama [City]*, and on ship board, and become much attached to him. I had found him possessed of many excellent qualities. Mr. Hart died at Weaver's camp just above Sacramento City about four weeks since. I did not know him but deeply sympathize with his afflicted wife with whom I am well acquainted.*

Young Depui is supposed to have been killed by the Indians. He, in company with Whitesides, Ray and others were in pursuit of some cattle the Indians had stolen from them, got lost and have never been heard of since. This was some five weeks ago and the impression at Sacramento is that they were murdered by the root diggers.

Mr. John Wharton lies very sick at Sacramento City; also a man from Dubuque County by the name of Heacock. Joseph Douglas, Edwin Coriell, Hadley and myself are at work together. We are doing moderately well. I met Mr. Reed from Cedar Rapids last Sunday and had a very pleasant chat with him. He informed me that Frank Charles was at Stockton.

You may expect me to start for home by the January steamer.

Yours, &c.
W.H. Merritt[76]

William H. Merritt to his brother George, November 2, 1849, from the Tuolumne River:

My Dear Brother,

A rainy day has at length made its appearance, the first since I arrived in the country, and being compelled to suspend operations, I have concluded to occupy the afternoon in writing to you. I have written you three letters since I arrived in the country, but being entirely cut off from all communication with San Francisco, I am unable to acknowledge the receipt of any answers if written and forwarded. Do not understand me that there is no communication with San Francisco from the mines, but merely that it is impossible to get letters unless you go in person to the office.

As an evidence of this, I have sent four times for letters and twice have I taken the trouble to write a letter to the postmaster politely and urgently requesting his honor to forward them, but to no purpose. I trust however, and indeed I have reason to know that you have not forgotten me, although were you to see me now in my old rusty and tattered garb, I cannot but think you would feel a strong disposition to pass me un-saluted were you so fortunate as to recognize me as an old citizen and friend. I am really what the old adage says, "fat, ragged, and saucy."

My health is good and has been since I arrived in the country, a circumstance quite fortunate in more respects than one. It only costs $12 per day if you deposit yourself in a California hospital, which is merely a canvas tent, and even at this enormous price there is no solicitation for custom. Men in the mines are as independent as wood-sawyers, and with a pick, pan and spoon in place of a saw, buck and file, resemble them very much. Without exaggeration California miners are the most wolfish and frightful pictures of mortality that I ever beheld, but like singed cats they are much better than they look.

Speaking of my own health reminds me that I must not forget to say a word as to the health of the miners. There is a good deal of sickness in the mines and I think in a great measure unnecessary. Scurvy and dysentery are the only complaints known here and I think three fourths of the mortality would be avoided if people would expose themselves less. I am aware that exposure to some extent is absolutely unavoidable; the newness of the country and the absence of building materials and the production of vegetables make it so.

...A large portion of the miners here make their diet almost entirely of pork and sea bread and this as a consequence produces scurvy, one of the most loathsome diseases on record. The doctors advise to the contrary, but the mass pay no regard to it. I suppose they think, as they have come all the way to California to be the makers of their fortunes, they must play doctor and be the curers of their own complaints. Well, poor fellows, they have come a great distance from home to die. No doubt they have left warm hearts behind to deplore their loss.

...But I must pass on to other subjects. Well my dear brother how would you and my friend Judge Dyer like to take a little turn with me at digging? Did you ever imagine what gold digging was? If not, let me ask you how you would like digging on the canal, quarrying rock under a broiling sun, or going into the cold water up to your waist before sunrise. Gold hunting is decidedly and emphatically the severest labor known to man. I have seen the ambitious Yankee, the hard-fisted Irishman, and the untiring German give way and abandon the mines with the benediction,—God save the man that's compelled to work in the mines—a very sensible and just benediction.

I first landed upon this river, but have since visited Wood's Diggings, Sullivan's, Sonorian Camp, Curtis' Creek, Mormon Gulch, Stanislaus River, Angels Creek, three branches of the Calivanus, McCalomy [Mokelumne] *and Mariposa. Among this number Woods, Sullivan's, Mormon Gulch, Sonorian Camp, Curtis' Creek, Angels camp and Mariposa are what is styled dry diggings where exclusive coarse gold is found.*

On the rivers the gold is fine. The river diggings are wet, but as a general thing pay much better than dry diggings. The manner of working is to dam out into the river and wash the dirt in the bed of the stream, but with the poor materials for damming, the holes require much bailing which is extreme hard labor.

The manner of working dry diggings is to sink down from 4 to 20 feet to the clay or slate rock and then drift. The work in dry diggings is much pleasanter than wet, but much more uncertain. If lucky, you get the gold in large quantities, but not more than one in fifty is successful. Hundreds and thousands here are not making their half ounce and thousands are discouraged and using every effort to get out of the country. God speed them in their efforts.

There is no doubt but there is large quantities of gold in this country, but the present condition of exorbitant prices for food and clothing render the chances for a fortune by digging entirely out of the question. Any other business is better than digging, but it requires a large capital to enter into business. It may appear strange when I tell you that I never was in a country where money was more plenty and yet where it is worth more than here. I perceive I must close. Express my tender regard for my sister.

Your brother,
Wm. H. Merritt[77]

William H. Merritt to unknown addressee, January 12, 1850, from San Francisco:

You are aware I was detained upon the Isthmus for nearly three months. This was very expensive, so that after paying my passage, ($150) I landed in San Francisco with but six shillings; not enough by ten shillings to procure a meal of victuals. As soon as I landed, I offered myself for service and obtained a job of turning grindstone for two hours at one dollar per hour. I soon found another job of repacking bottles of liquor which lasted two days at ten dollars a day. At the end of this service, I met a friend who voluntarily proffered me $100 which I accepted, and disgusted with my bad fortune, immediately started for the mines.

My bad luck appeared to follow me there, and while some, with moderate work, made out well, I labored incessantly and made but little. I returned from the mines the other day with seven hundred dollars which was obtained by the hardest labor man ever endured. It cost me $100 to reach this place from the mines, and after paying my debt, I had $500 left. This is a small amount for the city of San Francisco, but still I shall remain here or at Sacramento in preference to going to the mines again. Mining is decidedly the poorest business in the country.

I think I shall go to merchandising in the spring at Sacramento City. I have made arrangements with a friend with that view. It is an excellent business and many fortunes have been made.

It is said that Judge Hastings has made ten or fifteen thousand dollars in real estate since he arrived in the country. Large fortunes have been made in this business, and I think from the vast immigration that is reported to be coming, property will advance very much for a year to come.

I have met with P.B. Cornwall who has been in this country two years and made from 3 to 400,000 dollars. I am satisfied he is immensely rich. Business is rather dull in this place now but everyone anticipates a great rush in the spring.

Poor Joseph Hempstead we buried a few days since. Myself and Bothwell sat up with him the night before he died. He told me to say to his friends that he died happy and perfectly resigned; indeed he was anxious for the hour to arrive for some days previous to his death. In a post mortem examination it was found that he died of cancer of the stomach.

Bothwell is about to open a store. E. Coriell is here in no business. Thad Martin is here. Charles Miller is at work at his trade at $12 per day. F. Roberts is here waiting for the rainy season to pass over in order that he can go to the mines. A brother of L. Moloney is here, also Sam Cox and Maxwell. Maxwell has made about three thousand dollars. I understand that my friend Berry, Doctor Crane, James Fanning, William Abbe, and many other of our Dubuque and Iowa friends are at Sacramento.

…I have never passed a day since I left Dubuque that I have not been homesick, and I could pray with the fervency of a saint that my punishment was at an end. Yet my mission is unfinished, and while that remains, I must remain to suffer. Thousands are leaving disgusted with California, and thousands more of poor forlorn creatures would leave had they the means to get away. But this, thank God, is not my case. I have a "destiny to fulfill" and until that is accomplished or the means for accomplishing it are exhausted, I must remain.

Although I had but a few hours to reflect upon the importance of a trip to California, yet I flatter myself I fully considered the vast responsibility of the undertaking. True, I have not met with that success I anticipated; on the contrary, I have been pierced on every hand by the frowns of fortune. But I am determined not to give up the chase. I will not forsake my patron saint. I will still continue to worship at her crowded altar in the fond hope that by and by my devotion will be honored with a gracious smile.

…

Yours affectionately,
William H. Merritt[78]

Chapter 7

Overland to the Land of Gold

George Asbury Madeira and Family Cross the Plains

Going to California in 1852, as George Madeira did, was not as risky a venture as it had been for the earliest migrants in 1849. The trail was much better known, including the various cutoffs to take to shorten the journey and those to steer clear of to avoid disaster. Suppliers of goods and services had been developed, and bridges and ferries to cross streams large and small had been constructed.

The Native Americans had become less hostile. This was due in part to the loss of large numbers of their people to diseases to which they had no natural immunity—diseases that had been brought to them by the forty-niners. They also came to the realization that, given the vast numbers of migrants entering and traveling through their lands, resistance was futile.

Knowing of the improved conditions regarding the long trek, Madeira, a lawyer by profession in Dubuque, and his family made the decision to leave for California via the overland trail across the Great Plains. George was then fifty-one years of age; his wife, Susan, was forty-one. They took with them their three sons: Francis, twenty-one; Daniel, sixteen; and George, fifteen.

George, the father, had an abiding interest in development of railroads across Iowa and into the Plains states. Perhaps he had performed legal services for investors looking to make money in lines already in place east of the Mississippi or in constructing railroad bridges across the Mississippi and laying new track across Iowa. He does not speak of being an investor himself.

George, the son, became quite well known in his adult years in California for having established the first astronomical observatory in that state. He

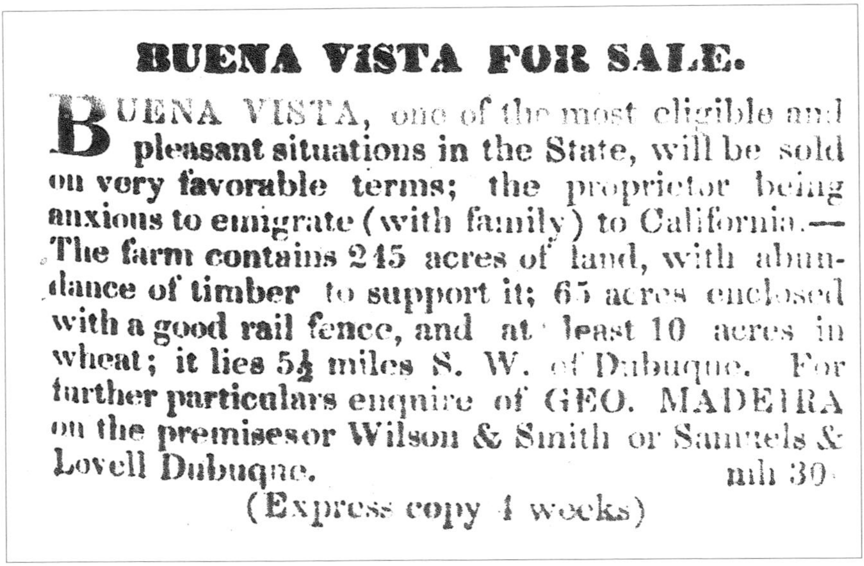

BUENA VISTA FOR SALE.

BUENA VISTA, one of the most eligible and pleasant situations in the State, will be sold on very favorable terms; the proprietor being anxious to emigrate (with family) to California.— The farm contains 245 acres of land, with abundance of timber to support it; 65 acres enclosed with a good rail fence, and at least 10 acres in wheat; it lies 5½ miles S. W. of Dubuque. For farther particulars enquire of GEO. MADEIRA on the premisesor Wilson & Smith or Samuels & Lovell Dubuque. mh 30

(Express copy 4 weeks)

Buena Vista for sale. Property of George Madeira. *Dubuque Tribune*, 1851.

is credited with having given inspiration to James Lick to use some of his wealth to build the famous Lick Observatory.

While many of the Dubuque California travelers returned to Dubuque to live out their lives, the Madeiras were not among them. George A. Madeira died in Carson City, Nevada, in 1865. His letters offer a graphic account of the two-thousand-mile walk across the West, providing a vicarious experience for today's reader.

George Madeira to A.P. Wood, March 12, 1852, from Kanesville:

> *Esteemed Friend,*
>
> *Agreeable to promise, and in accordance with an expressed wish of several of my Dubuque friends, I attempt a cursory account of our travels and observations of the country thus far on our way to California.*
>
> *The major part of the road from Dubuque to Kanesville is generally good, and can at small expense be made equal to any road found in the West. The accommodations for emigrants* [are] *very indifferent, and will remain so until other tenements are built and good wells of water are provided. The face of the country generally presents an undulating appearance, but not of so abrupt a character as to cause apprehension of its being impracticable for the building of railroads.*

...Kanesville is situated between two points of bluffs, about four miles from the steamboat landing upon the Missouri River; fronting the place is a beautiful bottom, in appearance like a prairie, containing an area of land of not less than thirty thousand acres, which in time will be covered with highly cultivated farms. There are now many good farms upon it generally owned by Mormons, who are now disposing of them or giving them away, as there are not purchasers for them, and they are bound to leave for Salt Lake, having been ordered so to do, under the pains and penalties of an exclusion from the Paradise of God if they do not heed or obey the mandate. Poor deluded creatures, they are to be pitied.

Hyde, Young, and other of the leaders of the Mormons, are mere sensualists, all justifying the crime of polygamy, which to them will eventually prove their destruction as such debased characters cannot be permitted to go unwhipped of justice; neither the common or statute law will allow of it. As the Mormons leave, good men will take their places, and soon this country will present an aspect of cheerfulness. There are now here many respectable persons who will greatly rejoice when the Mormons leave.

None of the Dubuque emigration have arrived as yet, but I expect some of them on tomorrow (Saturday). I have heard of several who are on their way. Friends Donnelly and Cannon, as also Rickey and Van Hagan, I expect them all tomorrow, they have had a time of it, as the roads are very muddy. I had good roads and have been here one week. I will write you again on our all getting together. As ever dear friend,

George Madeira

P.S. March 13th
I am with my family at Council Point on the Missouri River living in a tenement belonging to Col. Hardin, uncle of the lamented John J. Hardin of Buena Vista memory. The Colonel lives in the same yard, having as a companion a most worthy and intelligent lady, as also sons and daughters grown.

Since we have been here our dinner table has been consistently graced with wild geese and turkeys. They are so abundant that we can stand in the door of our dwelling and bring them down from their companions in flight at almost any time of the day. My family and myself are becoming rather tired of such good cheer fearing a _____ _____. At my present sitting, I have _____ _____ very large goose and six large Mallard ducks being prepared for a roast to be used by us on the Sabbath. This morning we had a roast Brant [?] *for breakfast. This will do to give you an idea how we live. More anon.*

G.M.[79]

George Madeira to A.P. Wood, April 27, 1852, from Kanesville:

> *Emigration to California and Oregon are setting in the way quite fast, I should say. There are now here some five thousand people en route for those places, some have already crossed the Missouri River, and are now wending their way to the land of gold; others will leave on the 29th and will continue to leave until all have crossed. I think the Dubuque Company will leave about the fourth of May.*
>
> *Corn, oats, and hay are plenty. I bought yesterday shelled corn in sacks at 25 cents per bushel, and oats in sacks at 28 cents. There is more corn and oats in this market than will be required for the emigrants. All our Dubuque friends here are well and in fine spirits; none who will not be well provided for the trip across the plains. It is the intention of us all to travel leisurely, and when we can with _____ to our teams; rest on the Sabbath day.*
>
> *...Just as we leave here I put off writing you a few lines, noting our organization, &c.; until then adieu.*
>
> *Sincerely your friend,*
> *Geo. Madeira*[80]

George Madeira to A.P. Wood, May 30, 1852, 525 miles from the Missouri River and 186 from Fort Laramie:

> *My Esteemed Friend,*
>
> *In resuming my journal I find I have to be very succinct as time is kept in constant requisition by calls for other purposes. None know or can realize (except those who have the experience of the emigrant across the Plains) how much depends upon a watchful care of the horses and oxen, which all careful companies give both day and night, knowing their very safety depends upon the preservation of their teams.*
>
> *The major part of the Dubuque emigrants left Kanesville (as you know) on the 5th inst., and were crossed over the Missouri River on the 8th at night. I having, with my family, crossed the preceding day. Since then we have been wending our way toward the land of gold, only stopping over Sabbath, which we are now doing at my present writing, and a lovely morning. We have good water at our tent doors, and most excellent pasturage for our horses and cattle. Our hour of preaching today is ten o'clock, the sermon to be by the Rev. Mr. Shafer. Rev. Mr. Hulbert preached last Sabbath.*
>
> *On the 11th, at the ferry on the Elkhorn, a new organization of the Dubuque Company took place, a large proportion having seceded, forming*

themselves into independent divisions, calling themselves "The Independent Dubuque Company, Nos. 1 and 2"; D.M. Morrison commanding the first, and A. Perrin, the second. The old company from which they seceded retaining the name of Rickey's Company.

Both the other Dubuque Companies have been behind us for the last two hundred miles, but Morrison's Company will lead us for a few days, as they have passed on, and will travel today. Perrin's Company is encamped about ten miles in our rear, and will remain in camp today. In passing along in my narrative, I cannot refrain from expressing my humble thanks to Almighty God, for his manifest beneficence in our preservation, as thus far we have been kept in health, and without the loss of either horses, oxen, or wagon.

I have just returned from hearing a most excellent and appropriate sermon by Rev. A.H. Shafer, one of our company; there were quite a number of emigrants, strangers, present, all of whom gave close attention to the sermon. Mr. George Eggleston and James Logan, with families, came up to our encampment just as we were about to assemble for preaching, and remained a few minutes, and then left for an encampment about seven miles ahead. They were all well and reported both Perrin's and Morrison's companies as all well, and stated that the Morrison train passed us last night. Messrs. Donnelan and Connel's families are in the Morrison Company, as also the Yates'.

In Capt. Thomas Rickey's Company are the following Dubuquers: J.P. Van Hagan, J.B. Van Hagan and families, myself and family, and William C. Abbay and family, all in good health and spirits, hoping in due time to reach our destined country. There [are] *a great many emigrants on their way to Oregon* [unreadable line] *at the upper ferry on the Missouri when we crossed, and I presume an equal number crossed at the other two ferries.*

At a ford about fifteen miles above Fort Kearney, several hundred teams crossed, and some two thousand teams had passed up the Platte on the south side. A very large emigration are on their way to the Pacific country; hundreds of families in company. I fear many are destined to great suffering, but pray God in his infinite mercy to extend his fostering care over them, averting disease and famine.

I have seen several new graves along the way, two of them contained the lifeless bodies of two heads of families, one of four children and the other of five, both brothers-in-law; they were from Missouri. The wives, after burying their husbands, left with their families for the state of Missouri. They contracted their disease on the south side of the Platte, which was aggravated in their efforts and exposure in crossing over the river.

I saw both of these families in their distressed condition, one while the corpse of the husband was laying in the tent; the body was wrapped in a winding sheet and interred without a coffin. Whatever sickness is found on this side of the Platte, I have no doubt was brought over from the other side of the river. The north side of the Platte never should be taken by the emigrant as it has proved heretofore most healthful. [sic]

I have not met with any who came by Kanesville who were sick.

...

Best respects to all friends.
Sincerely your friend and servant,
George Madeira[81]

George Madeira to A.P. Wood, June 17, 1852, from Rock Avenue, 630 miles west from Kanesville:[82]

My Esteemed Friend,

In resuming my journal up to within fifty miles of Fort Laramie (of which I have noted you,) I have to say we have met with many scenes and objects, calculated to interest not only the naturalist, the chemist, and the florist, but every other observing person; but not being a proficient in any of those sciences, I shall forego an attempt of an analytical description of anything presented to view.

The Sioux Indians along the Platte Valley are a fine specimen of nature's noblemen. Alexander, upon his admirable horse Bucephalus, never made a more imposing appearance than does a Sioux upon his well caparisoned horse; and the squaws make a no less interesting appearance; their costume is both rude and becoming, decorated with beads, wampum, porcupine quills, &c. They occupy the masculine position upon a horse, and are equally expert in horsemanship.

...All the Dubuquers are thus far on their way to California, well, and are getting along on their way more expeditiously than the emigrants generally. Morrison's and Cook's Company encamped immediately in our neighborhood last night. Donallen and Connell were with them; Perin, Eggleston, and Logan, with Strohl are several miles ahead. Captain Rickey's company contains four Dubuquers: Rev. W. Hulbert and lady, the Van Hagans and family, and self and family.

We expect to spend our Sabbath in the neighborhood of what is called the Devil's Gate. I hardly think any company on the road to the Pacific has been more highly favored by our Heavenly Father than this commanded by

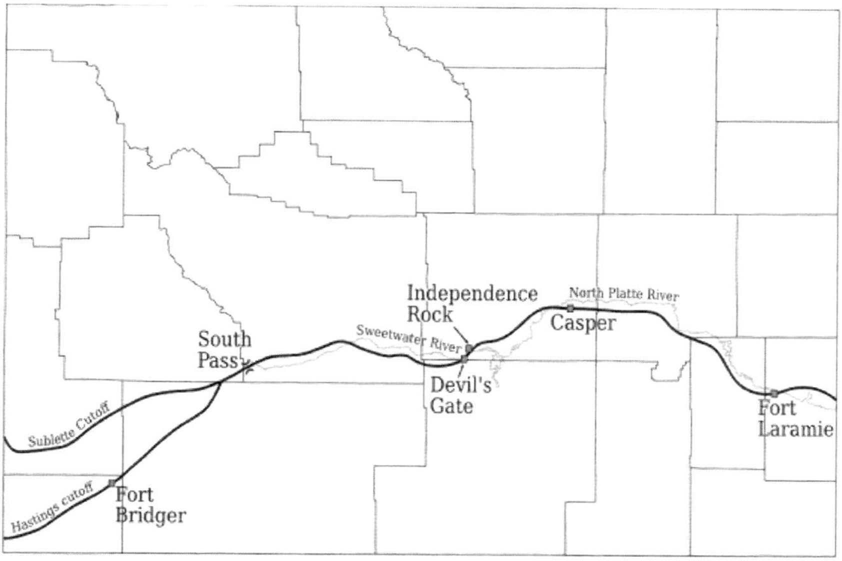

California Trail in Wyoming. *Courtesy of Tom Klein.*

our fellow townsman and friend, Captain Thomas Rickey. This company send their greetings to all their friends and acquaintances, hoping prosperity and happiness may attend them. Kindest regards of self and family to your lady and self.

Truly yours,
George Madeira[83]

George Madeira to A.P. Wood, June 19, 1852, from Devil's Gate, 704 miles from the Missouri River:

In my letter from Rock Avenue, I informed you it was the intention of our train to spend the Sabbath at Devil's Gate; I am happy to inform you we are now encamped in sight of this wonderful curiosity, where we expect to remain over Sunday. What makes this place such a wonderful curiosity is the passing of the Sweet Water River between perpendicular rocks four hundred feet high; width of river between rocks about fifty yards, the water rushing through its passage something like a cataract.[84]

The Independence Rock, situated on the road five miles east of Devil's Gate, is another curiosity, one upon which thousands of names have been cut, carved, and painted, among which names nearly every emigrant will recognize some with whom they have been intimate. I recognized a number written by esteemed friends from Dubuque, such as James Fanning, F.K. O'Farrel, the Woodwards, Roberts, Harts, and others. Our company dined at the base of this celebrated place. After dinner, self, wife, and others of the company ascended to the top, where we found many strangers promenading and regaling themselves upon its summit. A splendid prospect is had from this rock. The valley along which the road runs is seen both east and west for a considerable distance.

Whoever from Dubuque may visit this rock will not fail to recognize the familiar names of this year's emigration. The Rickeys, Hulberts, Van Hagans, Cooks, Morrisons, Myers, Perins, Egglestons, Donallans, Connors, Madeiras, and all of whom (except Perin and Eggleston who are in advance) are encamped in sight and will remain over Sabbath.

Not the least of the curiosities on this road are the Alkali Springs and Lake. Those situated five miles east of Independence Rock furnish the emigrants with all the saleratus they need. Our company, each family of our encampment, furnished themselves with abundance to do them to California. It is so plentiful that a person could shovel up a wagon load in an hour. This saleratus is an excellent substitute for soap.

...Should I not have another opportunity to write before we reach California, I will on my arrival there give you an account of our

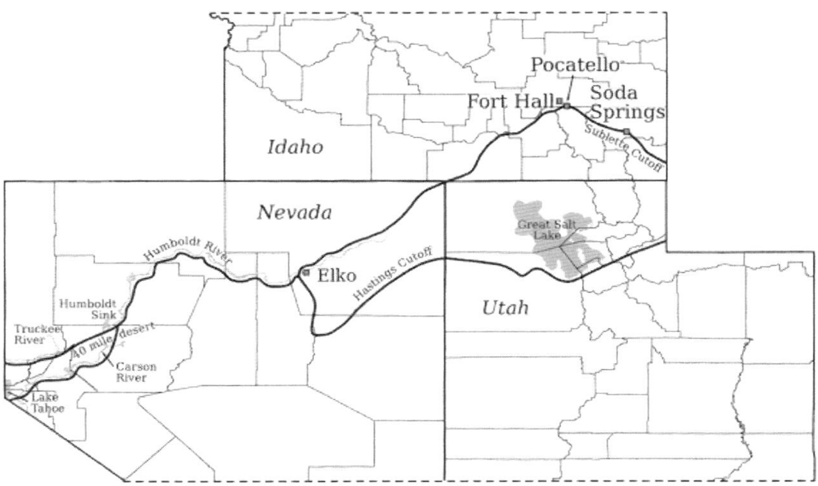

California Trail in Utah, Idaho and Nevada. *Courtesy of Tom Klein.*

further progress. I would have the relations and friends of the Dubuque emigrants know all are well, progressing expeditiously on their way to California.

Remember all kindly to friends and relations and accept our best wishes for prosperity and happiness.

Truly your friend,
George Madeira[85]

George Madeira to A.P. Wood, July 12, 1852, from Salt Lake City:

Esteemed Friend,

It is with pleasure I inform you of our arrival at this place, self and family came in on Thursday the 8th; friends Rickey, the Van Hagans, and Yates having taken the direct route to California. Since our arrival here Donnellen, Cook, Connell, Morrison, Connor, Col. Kane, McNair, Thomas Myers, Davis and some others of the Dubuquers have come in and are now encamped in our company, all well and much gratified at having taken this route.

My last letter to you was written from the Devil's Gate, giving you our travels generally. In our further progress onward to the Pacific, we have [had] *opportunities to observe many natural curiosities; among the most curious and astonishing is the Oil Spring. This spring affords to all who take the trouble to gather it, an excellent substitute for tar for wagons. Our road from Iowa to this place is superior to any road I have traveled in Iowa. But little difficulty presents itself in overcoming any of the hills.*

…The Salt Lake country is inviting to the agriculturist as his return in the product of the land is equal to that of our best land in Iowa. There are several thrifty farming settlements throughout this country; we have eighty miles of our road laying through a settlement of farmers. The best quality of flour can be purchased here at two dollars and seventy-five cents per 100 pounds. I purchased at that price.

The buildings here are adobe but look as well as painted brick houses—the color is lead; many of the buildings are rather elegant; several of the store-rooms are of equal size to those of P. Waples or Wilde. The Mormon Tabernacle (as it is called) is a very spacious adobe building calculated to seat twenty-five hundred persons. With aisles full it will contain some three thousand.

Today many of the Dubuquers attended church and heard Governor Young and Councilor Kimball press upon the consideration of their audience the Mormon doctrines asserting as a truth that Joseph Smith was a prophet and Hiram an apostle, and that every person would yet have to acknowledge it. The Mormons, in their intercourse with the gentiles, are polite and kind, evincing every disposition to be friendly to all. Was it not for their spiritual wife doctrine, I should feel no disposition to condemn them for their doctrinal belief. The best policy on the part of the general government to be pursued with the Mormons is to give them all the Territorial offices. I have understood recently that Governor Young would be continued in office; in this President Fillmore acts wisely.

My wife, Mrs. Connell and husband are waiting on me to accompany them to a remarkable spring just above our encampment. It is a hot sulphur one, rather warm for bathing purposes, but many avail themselves of the bath house to try the effect of it upon themselves. A short distance beyond there is another spring hot enough to boil eggs or meat. Friend Cook and self plunged ourselves into the fountain head of the warm spring which I found to be uncomfortably warm. I felt rather languid after having bathed.

For invalids the Salt Lake is considered to be the best to bathe in for the restoration of health. This lake produces an abundance of salt for the supply of the whole country; in the fall the people gather it up by wagon loads; it is very white and pure.

When in church, in looking out of one of the doors, I had a plain view of snow-capped mountains, which to me appeared like a freak of nature, but, my friend, it was stern reality—snow and no mistake. What a country this, to have snow in full view in July when in the valley the sun is rather uncomfortably warm.

Several of our Dubuque friends propose remaining here until spring. Mr. Cook's son and lady and child; Mr. Cline, William Marshall, Warren Robinson and William Stough. Their prospect of doing well is good. I had the pleasure of meeting with an old friend from Galena, Captain Hooper who has been residing here for upwards of two years and will remain here until next spring when he proposes to leave for California with intention to do business.

Mr. George Faiock has just requested to say that he is well and so far pleased with his journey.

At my writing this morning (Monday) we are preparing to renew our journey having purchased such articles as would make our journey more pleasant. We got butter at 15 cents per pound. On our arrival in

California you will hear from me. I hope yourself, family, and friends are all well. I hope for the time when I may take you all by the hand.

Sincerely, your friend,
George Madeira.[86]

George Madeira to A.P. Wood, November 23, 1852, from Volcano, California:

Esteemed Friend,

Time and distance have greatly intervened since I last wrote you. I am now in what is called the land of gold—where fortunes have been, and are now occasionally made in a very few days. Mining is not now as certain a business as formerly, but few, comparatively speaking realizing their expenditures, but hoped for developments of new discoveries of gold, may yet enable them to make their pile and return to their families and friends East. Many when they do leave will take the overland route, believing it to be the more healthy and economical way to get to the States.

...The [emigrants] *from Dubuque to this country are much scattered over the mines. Only Mr. John Perrin and his brother Isaac, beside my family in this neighborhood. I understand, however, that the Dubuquers are all well and generally doing pretty well. Should any from Dubuque think of coming to California in the spring, I would advise them to leave Council Bluffs crossing at the upper ferry about the 20th of April, taking no more flour or bread than what would be sufficient to do them to Salt Lake. I*

California Trail in California. *Courtesy of Tom Klein.*

will, however, write you again on this subject giving my advice in reference to an outfit.

This California is a wonderful country. The grass now just starting into life—and when Iowa and other states in the East are housed up on account of snow, we shall have the verdure and blossom of vegetation.

I am forced to bring my letter to a close as the post master has just informed me the mail will be closed in a few minutes. In my next I will write about both the Carson and Truckee route over the [Sierra] *Nevada Mountains. Family all well and best wishes to all our old friends hoping yet to see them.*

Truly yours,
George Madeira

P.S. Prater of Dubuque staid at my house last night and was well—wife in Iowa Valley at friend Rickey's.[87]

Chapter 8

A Woman's Perspective

The Experience of Mary Elmira Morrison

Mary Elmira Morrison was a strong-willed, independent woman who was put into the position of caring for her infant daughter and herself after her husband, Guy B. Morrison, departed for California in 1850. She was born on December 11, 1827, and at the time of his departure was only twenty-two years old.

The Morrison name was prominent in the Dubuque area beginning in the 1820s. At the age of eighteen, Guy was already a successful trader in Wisconsin at the village of Blue River on the Wisconsin River. Family lore has it that he was a personal friend of the Sauk Indian chief Black Hawk. In later years, he served as postmaster and also as treasurer of Dubuque County. He was a successful merchant in Dubuque but is supposed to have lost a fortune due to the collapse of the wildcat banks.

At the age of thirty-two, Guy left for the West on May 13, 1850, on his way to jumping-off at St. Joseph, Missouri, for the overland crossing of the Great Plains to California. During the crossing, the wagon train was destroyed, and only Morrison and one other man survived. Upon arrival in the gold fields, he first mined at Placerville and then at Oregon City, where he is said to have acquired "considerable wealth." (But see Guy's letter of January 19, 1854.) In August 1854, he became a judge of the justice court of Oregon Township, where he owned 320 acres of land near Oregon City.

In the meantime, while Guy was traipsing off to California, Mary Elmira was tasked to attend to various business activities and to raise their young daughter without his emotional support and with little financial assistance from him. At length, she concluded that he was unwilling to return to her in Iowa. Her letters reveal her longing for him and the shocking enterprise she undertook to settle the matter to her own satisfaction.

Letters from both Mary Elmira and Guy serve as a superb example of the dynamic that must have occurred many, many times between husbands and wives as the gold rush episode played out. Both parties, and any children as well, were accosted by pressures and strains that made life extremely trying. One can sense these pressures in the following letters.

The first letter from Guy is dated 1850, the time of his departure, but the second from him is four years later, after he had been gold mining in California. The letters of Mary Elmira serve to illuminate the initiative and strength of this woman. She died on August 28, 1881, and is buried in the Laytonville Cemetery, Mendocino County, California.[88]

Guy B. Morrison to his wife, Mary, May 10, 1850, from Indian Territory, seven miles from St. Joseph, Missouri:

> *Dear Mary,*
>
> *We crossed the Missouri on last Saturday and have been camped here since waiting for grass and recruiting our horses. We intend leaving for the plains on Monday next as we are tired of waiting for grass. We shall take a little feed with us. The prospect is now better for grass. Our horses are doing well at this place. The weather pleasant. I cannot say who we will go with as we have joined no company yet, but we will not start without being about 50 strong. There is a great many camped near us and a great many of them start on Monday, next. I wrote to you while at St. Jo. I hope you received the letter.*
>
> *I am going to church next Sunday. Perhaps for the last time for some time. I shall go to St. Jo's tomorrow and remain all night and attend mass early so as to return to the camp before night. I am going to confession and communion next Sunday morning.*
>
> *This perhaps is the last time you will hear from me for some time. But I will write to you by every chance. I see they have established a rail line from the sinks to St. Jo. and Weston.*[89] *I can write from there if not sooner. That place is beyond Mary's River.*
>
> *I tried my hand at washing yesterday. I am but a poor washer but hope to improve. I am a tolerable good cook. I am well pleased with my partner Henry. He is a good young man; upon the whole we are getting along very well. Uncle Sylvester is a good woodsman. We have all been very well since we left Dubuque. I believe I am better than I have been in sometime. I do not suffer with my back at all. I believe the trip will do me good.*

Robbins and wife are going with them. I am glad we got rid of him. He is not of much force. Priest I heard preach in St. Joseph on last Sunday week is on his way to California. He is only a few days ahead of us. I must come to a close. We are all busy fixing everything for the plains. Remember to pray for me as written by you in my portfolio.

Give my love to all and believe me as ever,

Your affectionate Husband
Guy B. Morrison[90]

Guy B. Morrison to his wife, Mary, January 19, 1854, from Dixon Ravine, California:[91]

Dear Mary,

Thy letter of Oct. 7 and 22 are received and I am glad to hear from thee and our dear little girl again. I also received the paper. I [am] *very much pleased to see a Dubuque one. It is the second one I have seen since I left home. I hope thee will continue to send it to me. I have nothing new to relate of importance.*

We have had thus far a very cold winter for this country. It is snowing now and the snow is a foot deep. It reminds me of Iowa. The oldest settlers never saw such a winter in this part of California. I have made nothing, but Edward is living and working with me. He had a blow up with Jesse and family. He says his sister M[ary?] *is an awful woman and I think so too. I hardly ever go to the ranch. I took dinner there Christmas. They sent for me and I had to go.*

Thee speaks, Dear Mary, of our long separation in serious terms and I am not astonished, but how in the name of God can I help it. I have tried hard to make money but can't help bad luck. Thee says I have been away over three years and have sent thee only $280 dollars. It is a fact but I have nothing to uproach myself of. I have sent thee all I could and that is all I could do.

John is living at Jesse['s]. *Uncle Sylvester lives about 3 miles from me. You say that thee will come to me in the Spring if I do not return. I do not know what to say to this. If thee enjoyed good health I would be glad to have thee with me in California if I should stay longer, but what are we to do with our dear child. It would not do to be without her, for her I must see if we live.*

Before much I am still in hopes of making money this winter. We have not had a favorable winter for mining to this time. I am in hopes the rain will last this spring so as to give us time to work out our claims.

It is hard, Dearest Mary, to be home. Thee certainly does not wish for me to return a poor man. How can I go back without some money? What could I do in the States a poor man after being in California? Nothing. If I am to be poor California is the country for me. The finger of scorn shall never be pointed at me in the States. I sincerely hope to make enough to go home with Ed this spring.

My health is poor and I fear I never will be as hardy as I have been. Excuse my writing, my hand is still. I can hardly hold my pen. Write to me often, and [send] *the papers. Kiss our dear child and give my love to all the family. I remain as ever, thy affectionate Husband,*

Guy B. Morrison

I do hope thee will send the likenesses spoke of in thy [letter]. *I should like to see the likenesses of all that is dear to me on earth. Oh God, may I live to see the originals.*

I am,
Guy[92]

Mary Elmira Morrison to friends, January 26, 1854, from the Atlantic Ocean:

Dear Friends,

The days have slowly passed since we left New York, and we are still gliding along on the solitary waist of waters, not having seen land since the first day out—and but two vessels, and then at a great distance. But, so far, we have met with no accidents and all, though we have had a very rough passage. Both crew and passengers are in good health, seasickness excepted. We have seven hundred passengers on board: lost one man overboard on the first night out; and have had two births since.

Cousin Grace and husband Lirgy, Uncle Amos, Cousin June and her son Marmaduke all accompanied me onboard and shed me all the kindness and attention it was possible, and I shall not forget them all. When we came onboard cousin Joseph Davis (Grace's husband) in trying to find the Captain, came across the engineer (or Chargé de affaires) who ______ knowing his business with the Captain was to place me under his care, kindly offered to take charge of me himself, and well has he fulfilled his promises. He has Anna with him the most part of the time in his room.

...We will reach San Juan[93] *in four days. From thence, across the Isthmus, and so on. I shall send this back by this boat, and when next I*

write it will be from Bidwells' Bar.[94] *Anna thinks the time long before she can see her papa. She often talks of home, and since on the vessel says she wonders how to get a good drink out of the well as for her sickness. I never suffered one day in my life, and Anna was sick enough, but as we were in no danger I do not mind it much.*

January 27th. We passed Cuba this morning—4 o'clock. I did not see it, as contrary to our expectation we did not stop. Anna has been very much broke out for some days, but not in the least degree sick with it. The doctor says this morning she has the chicken pox, but that he considers it a most fortunate circumstance as she will not now be troubled with prickly heat on the Isthmus, and she will be so liable to take any other disease. She is playing about this morning in great spirits. She was greatly delighted this morning at seeing a shark pass us, and at the sea birds and flying fish.

Sunday 29th. We yesterday stopped at Kingston, Jamaica and took in coal. I did not go on shore as the sun was so hot, as hot as the warmest day I ever saw in Iowa. The town is almost entirely inhabited by black people. They brought great quantities of fruit on board to sell to passengers, but I bought only a few oranges as I was afraid of the others.

Anna says tell the boys if she could see them she could tell them lots of funny things she has seen. She was much frightened yesterday by seeing the little black boys swimming about the boat like a parcel of frogs. And when the passengers would throw them a dime they would dive after and bring it up. One little fellow was not bigger than Anna. Some twenty dimes were thrown to him and he got them all.

I saw some of the most beautiful flowers I ever saw yesterday but they have a disagreeable sickening perfume. If Mrs. Dougherty makes any inquiry about my trip, tell her not to start with some gentleman as a protection and that the passage from New York to San Francisco by Nicaragua in the first cabin is $275, in first class saloon stateroom $250, second cabin $175.

Of the first and second I do not know what is the price but it looks like a dismal place. It makes the heart sick to look at a den of so much filth and dirt. I took my passage in the second cabin ($175) but had the good fortune to never have to spend one night there; the officers of the boat were kind enough to give me a place in the first cabin saloon.

We are obliged to pay 15 cents per pound for all baggage over 25 pounds in crossing the Isthmus; one hundred and 60 allowed free on the steamers. There are many persons onboard who have made the trip before who say the vessels are much better on the Pacific side, and that there is very little difference in the first and second cabins. The Sierra Nevada *is our vessel on the other side.*

January 31st. We will reach San Juan [River] *in an hour or two as it takes* [i.e., floats] *the small river steamer that takes us across the Isthmus. Give my love to all. We are in good health and grateful. Anna sends love to all of you. I must now get my letters into the purser's office before she closes.*

As ever, I am your affectionate daughter and sister,

M.E. Morrison[95]

Guy B. Morrison to his wife, Mary, February 2, 1854, from Dixon Ravine, California:

Dear Mary,

I received thy letters of Nov. 22nd and Dec. 6th today. I was very much disappointed in not getting the likeness thee promised to send me. I shall hope to receive them from thee. If thee knows how much I desire to see them thee would send them very soon. I am very anxious to see how my dear little girl and wife looks. Thee can better imagine my desire to see them than I can describe.

I do sincerely hope to make enough to go home this spring but God only knows how I will come out. Digging for gold is very uncertain. Some days I think I will make enough, and perhaps in a day or two almost discouraged—still hope. I must try for thy sake and little Anna. My feeling I cannot describe when I think of our long separation, but it is not my fault. I have done all in my power. Bad luck I cannot help and I have had enough of it in this country over.

Ed is still living with me. We are mining together. He has partly made up with Jesse's folks tho can't stray from them, although he says they have treated him badly. I hardly ever go to see them. The boys, some of them, have not treated me well. If I had left them sooner I would be better off. We have had a very cold winter for California and not a great deal of rain.

We had just got fairly work[ing] *on our diggings and hope they will pay us from 5 to 6 dollars per day, and perhaps—can't tell. I am so tired of this kind of life. I almost wish thee and little Anna was with me in my cabin. I cannot be happy until I see thee and our dear child. How in the name of God can I return to Dubuque without enough money to pay my debts and buy a home for us? California is the best for a poor man.*

Sell the lots in Dubuque if you can, as thee must be in want of money. I have made nothing since last winter. In fact, I was sick all summer and do not enjoy good health at present. The boys are all well and Jesse's family also, but they are as poor as I am. They have managed badly. Jesse's trip to Benicia cost them about 700 dollars. They remained only a short time.

Write often and give me all the news where old Brasher died. I think Lancaster came to California. I do not know what books thee speaks of in thy letter that Nightingale shared. Forgotten about it. Please explain in thy next letter. I have almost forgotten how to write. You must excuse, my hand is very heavy from hard work. Mining for gold is a hard and laborious kind of work. I wish I could find a small pile. I would gladly lay down the shovel and pick for I am very tired using them, particularly in bad luck. I do sincerely hope to make enough this winter to end our long separation.

I can assure thee Mary; I am doing all I can to get home again. I frequently cannot sleep after a hard day's work. Think of my hard fate—but God knows best. We should submit without a murmur, but it is hard to be apart so long. I shall use all exertion to make something this winter and leave as soon as I can for home. But, as I said, mining is very uncertain. I hope Mother Preston will succeed in her speculation. I will send thee some money as soon as I can.

Give my love to all and kiss little Anna for her papa. Do not forget the likeness.

Thy affectionate husband,
Guy B. Morrison[96]

Mary E. Morrison to her brother, February 1854, from Table Mountain Ranch, California:

Dear Brother,

We arrived here on the evening of the 19th. Anna well and hearty but myself just about worn out. When we left the Star of the West *we were in good health. We took the small boats and went up the San Juan River. We saw thousands of alligators, snakes and beautiful birds. We rode over the land a distance of twelve miles on a mule, as wild as a deer, with nothing but a rope around his neck. I took my carpet bag and _____ before me but like all the other women I was forced to ride both ways.*

When we got to the sea coast on the Pacific side the vessel could not come within a mile of the land. We were carried by the natives to the small boat that took us to the ship. In putting me in they let me drop in the water, and of course I was as wet as a drowned rat. I had put my trunks in the Wells & Fargo Express. When I left the Star of the West *it was to be delivered to me on the day we boarded the* Sierra Nevada, *but when we all got aboard, the trunk could not be found and I had to remain in my wet clothes until they dried on me, and by the time they were dry I was shaking with the Chagres fever. I never left my stateroom on Saturday. Up five minutes.*

After the first day out we sailed on the 6 of January being three days and four nights crossing the Isthmus [of Nicaragua]. *We had twelve hundred and twenty odd passengers on board. Had the small pox scare and scarlet fever and ship fever. Lost one man with small pox, the only death we had. We had an excellent doctor on board to which circumstance most likely I owe my life.*

We landed in San Francisco on the 16th. I saw Murry Morrison's advertisement in a San Francisco paper on the vessel. He is a broker. I sent for him and he met me not like a stranger, but had he been my brother he could not [have] *paid me more attention and been more kind. I found Guy in tolerable health. I took him by surprise as he had not got my letters, got them yesterday. Guy, John and Ed are building a house down at the cabin. They will not let me go down until they get things fixed. But I must now conclude with love to all in which Guy and Anna join.*

I am as ever, thy affectionate Sister
Mary E. Morrison

P.S. I shall expect some long letters soon. Tell Jesse to write and all of you write. Guy and Ed's claim is good, from $5 to $8 dollars apiece, a day. I will write soon and give some description of the country and diggings. I shall get one hundred and seventy-five dollars for my trunks.[97]

Mary E. Morrison to her mother, Anna W. Preston, March 17, 1854, from Buffalo Gulch, California:

Dear Mother,

I must commence my letter with excuses for not writing to you before, but perhaps it will be best to condense them into the least possible space and say I was sick for the first three weeks [with] *Chagres fever, and since then have been so busy as to be unable to write in the daytime and too tired to write at night. I wrote to Silvester as soon as I reached the ranch but was so unwell and so bewildered with fatigue and joy at meeting poor Dear Guy, that I do not remember what I wrote, so I will go back to the time I left the Isthmus.*

When we got on board the Sierra Nevada *we were in excellent health, but I got wet and was unable to find my trunk to get dry clothes, and thus got a severe cold which settled in Chagres fever, with which I was confined to my stateroom during the rest of our passage. Anna was well during our entire voyage and has continued to enjoy the same since our arrival.*

When we arrived in San Francisco I was without friends and my trunk was gone. The express agent paid me twenty-five dollars on my trunk, and

I sent for Murry Morrison (one of Uncle Robert's sons) who let me have fifty dollars; insisted on my taking more, which, of course, I did not do. He paid me every attention as also did his wife. A brother and sister could not have done more.

I stayed but one day in San Francisco, took a steamer from that place to Sacramento, thence on to Marysville. From there I took the stage to Bidwell's Bar, but when within about five miles of that place, and twenty from Marysville, at a place called Garden Ranch where the passengers took dinner, I learned I was nearer to Guy from that place than I would be at Bidwell's.

So, after some delay and difficulty, I procured a couple horses and a man to take me ten miles to the Table Mountains. We got to Jesse's about an hour after dark; but Guy and Ed were not at the ranch. Jesse went after them.

Guy was so surprised he would not believe I had come and he could not speak when he met me. I found him but little changed, and at first I thought in tolerable health, but have found since that he is in a very dangerous state of health. His lungs are much affected but for the most of the time he is able to work, or at least tries to make me believe he is.

We stayed at Jesse's a week while Guy and Ed were building a house. They would not let me come to their old cabin. We have an excellent house for this part of the country—a good clapboard roof, a plank floor. The house is lined with canvass. The two ends only are boarded in. I have one of the best cooking stoves I ever cooked by, a washboard and flat iron; also three pieces of household furniture not often met with in the mountains.

Guy and Ed are, for the present, working out a claim with John and Jesse, but as soon as the water dries up, will work just below our door. They average about $5 per day at their present claim. Ed boards with us.

We talk some of going down to the river during the summer season, it is only five miles. Guy will let out his part of the ravine on the shares, and he will help. We do the cooking and take about thirty boarders at ten dollars per week. Ed will also let out his share and pack provisions from Marysville so it will be of but little expense to get the provisions which otherwise would be a great tax. I shall know by the next time I write if we will carry this plan out.

John, Jesse, and the Doctor, together with loss of time, have nearly broke Guy and Ed, but now their eyes are once opened they will not be so easily gulled again. It will not be more than two years until we see you if Guy's health will permit us. We will, I think, go to Southern California next winter and see what good that will do him. This summer it will be warm enough here. Anna and I are getting as fat as pigs. I have not felt so well since I was thirteen years old as I now do.

Last week I was away from home four days. Was with a woman that was confined; Anna was with me. I got five dollars a day. She had a doctor sent for before she heard of me. Otherwise I should have made fifty. I am very sorry I lost my doctor book and medicine as they would be valuable to me now. I have two patients engaged; one about five, the other about seven miles from here which [when] *I am sent for will be fifty dollars.*

I expect thee will think I am in a fare way to make money. If I keep my health I will make money. A woman can do better than a man can. Had Guy sent for me one year ago, we could have now been worth a good many thousand dollars. However, it is not too late yet. We will send thee a draft as soon as Ed goes to Maryville. I think I shall go with him.

...Tell Jesse and Silvester to write often. I am looking for one daily.

Goodbye Dear Mother
M.E. Morrison[98]

Chapter 9

After the Dust Had Settled

The news that was to transform the nation struck Dubuque in late 1849—gold had been discovered in far-off California. Some writers believe that the exodus of people leaving for the California gold fields at mid-century caused great economic devastation in smaller communities throughout the States. The effect on Dubuque, however, a relatively large community, cannot be classified as devastation and was surely only short term.

Businessmen erected new, substantial brick buildings and new hotels, while other individuals built private homes. The city fathers turned their attention to the harbor to see to the adequate landing of steamboats and the economic benefits to be derived from this improvement. City revenue so greatly increased that it nearly wiped out the indebtedness previously undertaken.

The belief that the number of lead miners leaving for the gold fields was so great that whole communities were left deserted certainly did not apply to Dubuque. It is true that some left for California to "get their share of the rocks"; nevertheless, lead mining persisted in the region for many years, even into the twentieth century, though at continually declining rates from its peak in 1847.[99] Factors other than the gold rush coincided with that event and were perhaps more important in the declining lead industry. Primary among these would have been that the most easily obtainable lead had been raised and the underground workings were then down to the water level, requiring expensive equipment to drain them.

Chandler Childs, the editor of the 1880 *History of Dubuque County*, provides his assessment of the impact of those joining the California travelers on the city. He states that those who ventured into that comparatively undiscovered land numbered fully five hundred, mostly young men, among whom were farmers, miners, clerks, merchants and capitalists. Even though some left for California to join the gold rush, nevertheless, the incoming tide of new immigrants seeking a new life or farmlands in the fertile soils of Iowa more than made up for this population loss.

In examining population growth for the city, county and state, one sees that the number of immigrants substantially outpaced the number of those who left to participate in the gold rush. In fact, there were likely more who remained only a short time in Dubuque before leaving to take up farming than to pursue mining in a far-off land called California.

One wonders how the laborers in the lead mines would have been making sufficient money to afford to put together the kit necessary to make the journey to California, estimated at $300 for the necessary basics, in addition to another $300 in cash required to pay for ferry crossings and living costs until income-generating activities could be established. This was also true of hourly workers and day laborers in general.

For those who had done well enough to own property, the option of mortgaging it was available. Undoubtedly, there were those who chose this method in the full expectation that the wealth they would accumulate in California would erase the amount due on the mortgage when they returned.

The lead mine owners in the Dubuque area, on the other hand, were still making money through these years and would have had much less of an imperative to succumb to the "fever." People who had assets, such as business and professional men, and who were already driven by the entrepreneurial spirit sometimes saw an even greater opportunity in California and so gave in and took off for the gold fields. A report about a group forming up in adjacent Jackson County attests to this: it was "composed of some of our best citizens…many of them possessed of a good share of this world's riches."[100]

Another category of forty-niner was the farmer. These folks already had skills that would be useful on the trail and, perhaps more importantly, already had many of the items needed for the journey, especially the more costly ones—horses or oxen, wagon and rifle. They, likewise, were represented among the forty-niners.

While perhaps five hundred Dubuquers migrated to California during this gold rush period, there was a larger number arriving in Dubuque and Iowa in general.[101] So the impact on Dubuque made by those leaving was likely minimal, barely noticed and surely only short term.

Indeed, a certain amount of economic dynamism would have been generated by the activities associated with preparations to leave for California. Businesses, real estate and homes would have been sold by some just to raise the money for the journey. Further, supplies and equipment needed for the overland trip had to be purchased. Thus, even though numbers of people were going to California, their leaving was not a total negative for the local economy.

In actuality, the loss of local citizens to the gold rush cut both ways. There would have been empty storefronts on Main Street, professional offices would have been deserted and homes and farms would have been sold to raise money for the journey. Many of these sales were made at reduced prices, providing opportunities for newly arriving immigrants. The supply of money in Dubuque was taken to a reduced level, a fair amount having left with those who were California bound. The losses to the community were noticeable enough that A.P. Wood, the editor of the *Tribune*, wrote a booster column encouraging immigrants to consider Dubuque, where a goodly supply of real estate was available and the outlook for farming was most promising.

It must be remembered, too, that at this same time many migrants were arriving in Dubuque, adding their dollars to the growing economy. John Steele records his arrival in Dubuque as he was on his way to California on April 1, 1850: "Early this morning, we crossed the Mississippi on a small horse ferry to Dubuque. As the rain continued and the roads were deep with mud, the wagons were hauled into a yard within the city and we took shelter at a hotel. Crowds of California emigrants thronged the city."

Another forty-niner, Byron McKinstry, a schoolteacher traveling with the Upper Mississippi Ox Company from McHenry County, Illinois, also attests to the increased level of economic activity. After paying $2.75 to be ferried across the river, he arrived in Dubuque, estimating the population to be about three thousand, and "every one wide awake to make something off the Californians." His party put up at the Farmer's Home hotel, adding to the revenues of the hospitality business in the thriving community.

The spring tide of migrants was strong in 1849 and 1850, when the *Miners' Express* reported thirty-two teams crossing at Dubuque on March 23 and averaging perhaps twenty per day at this early time in the season. Many forty-niners from Wisconsin and northern Illinois crossed the Mississippi at Dubuque because they knew ferry service was available there.[102] Whether due to the delays in waiting to be ferried across or due to the perception that the cost was excessive, tensions sometimes exploded into physical encounters. Lucius Fairchild records in April 1849 that an acquaintance of his, a Mr. Riner, "had trouble with the ferryman and blows followed words. The ferryman hit him on the head with a stone and very nearly killed him, but, although it is a very bad bruise, he will recover after awhile."

Robbery, too, was part of the local scene with so many comings and goings. A trunk belonging to Fred Roberts was broken into and items stolen. The thief is supposed to have been in a party from Milwaukee staying overnight at the same hotel as Roberts and had left the next day down the Military Road to California. Roberts, it is reported, expected to head for California later in the spring "but would prefer that his watch and money should not get so much the start of him."

For the smaller communities surrounding Dubuque, a negative economic impact was more likely. Given the great costs associated with a journey to the gold fields, one has to presume that unknown numbers of forty-niners mortgaged or sold their businesses and/or homes or borrowed from banks, families, in-laws, friends and anyone else who would bankroll them. The resultant loss of these dollars to the local economies, especially in the smaller communities around Dubuque, would have been substantial. The impact on Dubuque itself, because of the constant flow of immigrants to and through the city and the resultant higher level of economic activity, would have been less noticeable.

The historical record is replete with indications that the discovery of gold at Sutter's Mill was the catalyst for a massive departure of the lead miners of the Upper Mississippi Lead District to the gold fields of California. It also posits the consequent decimating of the lead-mining activity in this region. Indeed, many miners around the nearby Galena area in Illinois joined the gold rush, resulting in little building going on there in 1850 and 1851.

In adjacent Wisconsin, sixty wagons left on one day from Mineral Point, and a total of seven hundred people, a seemingly exaggerated number, reportedly left from that small town. Reports indicate that Iowa also lost lead miners to the gold rush, supposedly leaving the mines all but deserted and taking many years for the industry to recover. And finally, for Dubuque itself, we find a report that declares the lead-mining production declined owing to the superior attractions of the gold fields of California.[103]

Yet the level of economic activity in 1849, the first year of the gold rush for Dubuque, was fairly robust. During that year, over eighty brick buildings went up, many of them substantial and at great expense. Money must have been plentiful and prospects bright. But, painting a different picture for the following year, Childs states that building of any consequence was lacking in 1850 and that "dullness reigned supreme." With the arrival of springtime, the business associated with the gold rush picked up but then suffered a relapse and stopped completely in the fall. In contrast to this negative perspective, a contemporary account likely provides a more accurate assessment: "The city of Dubuque is literally filled. There has been a greater call for houses within the last two months than ever before known. Houses are being finished every day, but are all engaged long before they are complete. More are building."[104]

While these statements attest to the widespread thinking that lead mining all but ceased with the mania to get to the gold fields, it is certainly true that lead mining persisted in the region for many years, though at continually declining rates from the mid-century levels.

The impact of the out-migration had only very short-term consequences, as Childs testifies:

> *The California wave had spent its force with the close of 1850, and, in the following spring, the city was granted another lease of life, so to speak. The land office was located at Dubuque. Emigration was resumed…new-comers entered lands and took to farming, invested in business ventures and added a new impetus to agriculture and trade…In short, the city and county commenced to fill up again, real estate appreciated rapidly in value, new buildings were erected, societies, banks and associations were organized…This influx and its sequences revived business, and so liberal were the daily accessions to the population, that, on almost every night during the ensuing months, there was scarcely a house in the city but what entertained travelers and prospectors.*[105]

The city was clearly on the cusp of a tremendous growth spurt, as can be determined by contemporary reports, and optimism ruled the day. Not yet seen on the horizon was the looming Panic of 1857. In 1854, the city erected 332 buildings, indicating that Dubuque was still prospering and growing. Another indicator of the dynamism of Dubuque at this time is the use of the Dubuque ferry. Timothy Fanning, S.L. Gregoire and Charles Bogy operated ferryboats in 1851. Bogy began operating a steam-powered ferry in April 1852, with excellent results: "She is doing the most rushing business of the season. She is puffing and blowing all the time. She is a perfect godsend to California emigrants. If the number of wagons that she brings across in a day had to abide the tardiness of the old-fashioned horse boat, they would not reach this side in a week."[106]

Additionally, imports to Dubuque increased from 32,007 tons in 1853 to 97,633 tons in 1854, with exports from the city increasing in 1853 from 7,482 to 11,736 in the following year.

Dubuque's experience as a nineteenth-century mining town encountering a mineral rush elsewhere is very uncommon. A large number of communities experiencing population losses in similar situations declined, with many disappearing altogether, as the next big strike became known. The list of examples is long and includes locales in Georgia, Missouri, the Upper Mississippi River Lead District, Michigan's Upper Peninsula, California, Colorado, South Dakota's Black Hills, Montana and others. The boom and bust cycle remained the same for lead, copper, gold and silver.

Dubuque, on the other hand, possessed advantages held by few of its mining peers. At a time when almost all pioneers farmed and most miners did some subsistence agriculture, Dubuque was surrounded by tens of thousands of acres of unclaimed, available, high-quality farmland. At that time, Dubuque was the largest community on the Upper Mississippi River north of Davenport, when the river was the major north–south highway for the region. At the same time, it was equally well placed on the emerging east–west transportation axis to provision forty-niners and other westering pioneers. Further, Dubuque's economy was large and flexible enough to absorb those who saw opportunity not in digging lead but in buying into existing businesses or creating new ones.

The loss to the gold fields of perhaps one-quarter of its adult workforce had only temporary effect on the local economy, and the income and population were quickly replaced by new businesses and an increasing influx of immigrants. Ultimately, the impact of the California gold rush on the frontier river city of Dubuque was minimal and of short duration only.

Those leaving for California came from a wide variety of occupations, professional men, merchants, miners, city politicians, farmers and day laborers among them. Women, too, joined the trek west, at times with young children and infants.

The lure of gold persuaded some of the lead miners to abandon lead mining in favor of gold mining, but the decline in lead mining was not caused *solely* by the gold rush. City population continued to increase apace, mirroring the population increases achieved by Dubuque County and the state of Iowa.

Some wealth was generated for the city due to the economic activity associated with locals and transients preparing to leave for the gold fields. Some additional wealth came back into the city by the returning Dubuque forty-niners, but the amount was not transformative. For most Dubuquers, the gold-mining episode was one that brought not riches but perhaps a more valuable commodity: wisdom.

Chapter 10

"Sorry, Elmira...Not So Much"

Did they strike it rich? The simple and direct answer is, "Not so much." The successes or failures of most of the Dubuque forty-niners are not known, there being no documentation providing evidence. One concludes that most were failures or only modest successes. Remember, however, that even if a Dubuque gold seeker came back with "merely" $300, a pittance in the gold fields economy, it represented approximately a full year's income in the Dubuque economy.[107]

The local newspapers show almost no reporting regarding failures and even less regarding monetary success. Then, as now, except for public employees today, such reporting was considered tactless. So the letters of the forty-niners themselves become the source documents for what little information there is in this regard.

Among the few known successes was William Bothwell, who wrote that he had not yet engaged in "regular" business in California, but in the meantime, he did a little dealing in town lots. In one case, he bought a lot on a Friday and on the following Tuesday refused an offer giving him a $1,900 profit—in only four days! He sold one of his houses, one presumes at a handsome profit, and the other one he was renting for $2,400 per year. These transactions were achieved from September to December 1849 at a cost of $1,400.

In the case of John Coffee, his wife and family, success came the old-fashioned way—through hard work. In October 1849, he was digging gold and getting about ten to eighteen dollars per day. His wife and daughter were taking in wash and making twelve to eighteen dollars a day. This was when twelve to fifteen dollars per day was an average daily income for finding gold.

Stephen C. Langworthy engaged in mercantile pursuits in California and "made a fortune" before returning in 1856. In a letter home, he tells of two of his companions who were doing quite well. The first, S.G. Williams, made $2,000 in one of those collateral enterprises that frequently provided more income than gold mining. He was a barkeeper on a boat running up to Stockton. The second, Joseph, the son of Captain George O. Karrick, returned to Dubuque with $6,000, quite a handsome sum. We know also of a man identified only as Maxwell who had made $3,000 by January 1850.

William H. Merritt, writing in November 1849, had returned from the mines to San Francisco with $700. We don't know, however, how much it cost him to get that. He traveled via the isthmus and encountered a long delay there waiting for transportation to San Francisco, thus adding to his initial expenses. Upon arriving in California, he then had to put together equipment and supplies, estimated at $300, to get to the gold fields. Since he later returned to Dubuque, he also had the expense of the return trip. The net return on his investment could have been quite modest.

Augustus Coriell had about $2,500, a tidy sum that was to do him no good. He made the mistake of trying to extinguish a lighted cigar in a powder keg by throwing water on it. His unfortunate attempt transported him to his eternal reward.

Peter Summers is said to have "bettered" himself; a man named Fulweiler was "more than ordinarily successful"; and another by the name of John Smith was reported to be "unusually successful."

Unfortunately, the evidence for Dubuquers who fell into the successful category is very slim indeed. Most would likely have been pleased to have made enough to repay the investment in putting together the wherewithal to get to California and to pay their way back to "the good life" in Dubuque.

One must evaluate all of the above dollar sums in light of their current values. Thus, the value of $1.00 in 1849 is equal to $28.80 in 2010, the most recent year provided by conversion tables. In the end, one is reminded of the old dictum of the surest way to come out with a million in any risky venture: "Start with two million."

Aside from those who had success in California, there were many more who came back empty-handed and even lost money given the expenditures involved in making the trip. We know that S.M. Hammonds returned to Dubuque after only one year and reported an experience "anything but flattering." W.S. Gilliam described how he felt about his California experience: "disgusted." There may have been some Dubuquers who made a moderate success and were unwilling to let others know about it, but it is probable that most were poorer, if wiser, for the experience.

In addition to those who were unsuccessful in returning with untold wealth or any wealth at all, there were those who paid the ultimate price and did not even get to return to family and friends, having died during their efforts to strike it rich.

Dubuquers fared about as well as immigrants to California from points across the States and around the globe. A few can be said to have done well or moderately well, but most had nothing to show for the effort except having "seen the elephant." Over time, they did accrue one other benefit. The local community held them in high regard—they were forty-niners.

Chapter 11

The Dubuque Forty-Niners

The names of the identified Dubuquers were retrieved from a variety of sources; Childs's *Dubuque: Frontier River City*, the Dubuque *Miners' Express* and the *Dubuque Tribune* are three such examples. Aside from knowing the names of the published letter writers, many additional names were acquired from the contents of those letters.

It is not at all clear that all of the people in the listing were actually Dubuque residents. Undoubtedly, some were from the surrounding region, had heard of a company leaving from Dubuque and decided to sign on with that group. It is also possible that some were only very recent arrivals to the city and decided to just continue moving on west without ever putting down roots in Dubuque.

Another difficulty is exemplified by the cases where only the surname is provided without the given name. An example of this is Kennedy (no given name) and then another entry identified as Kennedy, T.K. Are these the same person, or are they two different people? There is insufficient evidence to answer this question. Further, similar but different spellings complicate the matter. Are DePui and Dupui the same, or Edward and Edwin Coriell? And finally, is R. Reynolds either Richard or Robert Reynolds, or someone altogether different, named perhaps Raymond?

Whatever the case, some 298 names have been identified, these being mostly men, with only a few women and children named. The unnamed women and children would bring the total to over 300. One writer noted that a certain company was composed of a number of men "unburdened by wives and children." Was this a moment of chauvinism or just a clear-eyed recognition of the rigors and dangers to be faced on the journey?

The names on the next pages are of those known to have left Dubuque for the California gold fields or joined Dubuque companies along the way. Given that some three hundred have been identified by name, it is clear that the grand total leaving from Dubuque is a greater number. The number five hundred has appeared in the historical record, and that appears to be a reasonable estimate.[108]

It has not been possible to acquire detailed information on each of the migrants from Dubuque. Certain facts, however, can be ascertained from what is known. The average age of the travelers has been computed at 30.3 from a cohort of eighty-one known ages at the time of their travels. This matches the expectation that most of the Dubuque forty-niners would have been relatively young. Sixteen-year-olds were included in the calculation, but not younger children.

There were nineteen who fell into the category of businessmen or merchants. Among these were a baker, a confectioner, two selling dry goods, a druggist, a grocer and five simply identified as merchants.

There were thirty-six who had been associated with lead mining—raising or smelting. It is not surprising that this is the largest single occupation identified, since the economy of Dubuque up to 1849 was based initially and primarily on the lead-mining industry. It is rather remarkable that this number is not perhaps a larger percentage of the total number of those leaving for the gold fields. One must also take into account that some of these thirty-six had been associated with mining in the past but were no longer involved, and some were only part-time lead miners, a not unusual occurrence.

Seven doctors were known to have joined in the exodus from Dubuque, along with five attorneys. The state of professional certification at that time must be taken into account—who could call themselves a doctor or attorney?

Fifteen of the travelers were associated with city or county government, and eight were farmers. The number of women and children making the trip was quite small.

From the list of forty-niners identified, we have not been able to ascertain which route was used by all. But of those for whom we do know which route was chosen, most traveled overland on what became known as the California Trail. Fifteen crossed the Isthmus of Panama, two crossed the Isthmus of Nicaragua and one sailed 'round the Horn.[109]

The following list identifies the Dubuque forty-niners. Age at the time of leaving for California and occupation have been indicated when known. Names are recorded as they appear in original sources. A few of the names have been found in Randy Brown's *Historic Inscriptions on Western Emigrant Trails.*

Aldrich, J.B.	Age 31. Was a farmer in Wyoming, Iowa, in 1870 census.
Allen, Edmund M.	Age 25. Had a brother in Sacramento.
Allen, W.G.	
Alverson, L.M.	Traveled via the Isthmus.
Anderson, A.	Went to California in 1852.
Anderson, R.O.C.	President of Dubuque Mechanics Institute.
Andrew, William	
Anson, Washington	Age 32. Stonemason. Dub. Emigrating Co., 1852.
Archer, Aaron	Age 16. Had engaged in lead mining. Younger brother of the twins John and William Archer. Back in Dubuque in 1855.
Archer, John M.	Age 26. Had engaged in lead mining. Three brothers went to California in 1851 or 1852. Died in California in 1864.
Archer, William O.	Age 26. Had engaged in lead mining.
Bahl, Andrew	Age 17. Went to California in 1850 and was back in 1853. Probably a farmer.
Bemas, John	
Bennett, A.L.	Took his family on the trip.
Berg, Leonard	Age 26. He was a baker. Was in California for about 18 months.
Berry, J.V.	Took his wife and three children, one of whom was named Harry. Was a practicing attorney in Dubuque.
Blackwell, J.Y.	
Blake, Charles	Died at Sacramento.
Boone, J.T.	Was a doctor. Had been appointed to Dubuque Board of Health for cholera concerns. Appointed as surgeon of the company for the Dubuque Emigrating Association.
Booth, Edmund	Deaf, blind in one eye, limited speech, but otherwise fully competent. He and one other man went to California from Anamosa, Jones County. Traveled for a time with a Dubuque team.

Borchard, J. Christian	Farmer. Overland in 1849 with his wife Elizabeth and infant son J. Edward Borchard.
Bothwell, William	Age 40. Traveled around the Horn. Engaged in real estate in San Francisco. Made thousands in 1849. Back in Dubuque in 1855. He cultivated a large orchard at the corner of Grandview Ave. and Dodge St.
Boyles, Henry	
Bradstreet, Samuel Y.	Went overland. Took 112 days to Sacramento. Carpenter, builder, contractor. Married in Dubuque in 1855.
Branstreeter, William	
Buckman, M.J.	Came to Dubuque at an early day. Was joined a few years later by his sons C.C. and W.A. Went via the Isthmus. Mining and merchandising at Spanish Dry Diggings.
Burg, Leonard	
Burton, M.J.	Died on the *Golden Gate* bound for San Francisco in 1853.
Byers, William N.	Age 21. Later to become a famous Denver newspaperman and founder of the *Rocky Mountain News*. Joined a Dubuque company at Kanesville but then another company heading for Oregon.
Cannon	Dubuque Emigrating Co. in 1852.
Casteel, F.L.	
Casteel, J.S.	
Chalmers, M.	His wife accompanied him to California.
Chambers, Mr.	
Chapman, Theoph.	
Charles, Frank	
Childs, Thomas C.	Storekeeper in the city (Sacramento?).
Churchman, James	Lawyer.
Clapman, Theophilus	
Clark, A.V.	

Clark, Jonathan	Doctor. Left from Maquoketa. Many in the Dubuque area "acquainted with the writer," as he was a former Dubuque resident.
Clark, Terry	
Clark, William A.	
Cline	
Coates, William	Age 19. Went to California in 1852. Was a capitalist.
Coburn	
Coffee, John	Took his wife and six children on the trip. Dubuque County marshal in 1846. Ran the Iowa House hotel in Dubuque. Made $10–18 per day in the gold fields.
Connell, Samuel	Age 32. Carpenter. Dubuque Emigrating Co., 1852.
Connors	There were two Connors names inscribed on Independence Rock.
Connors	
Conron, W.	Went with the San Francisco Company out of Jackson County, Iowa.
Cook	Cook names inscribed on Independence Rock.
Cook, Lines C.	Dubuque Emigrating Co., 1852. Civil War casualty with 12th Iowa Infantry.
Cooke, Edward	Age 7.
Cooke, Eva Anna (Lily)	Age 10.
Cooke, John Richards	Age 21 when he went to California in 1852.
Cooke, Lucy (Rutledge)	Age 24 when she went to California in 1852. Sarah, her baby, born in Dubuque in 1851, also made the California trek.
Cooke, Richard	Age 5.
Cooke, Sarah	Infant daughter of Lucy Rutledge Cooke.
Cooke, Sarah	Age 44. Wife of William Cooke. Mother of William Sutton, John Richards, Thomas W., Eva Anna (Lily), Edward and Richard. Went to California in 1852.
Cooke, Thomas W.	Age 19 when he went to California in 1852.

Cooke, William, Sr.	Age 49. Clerk. Secretary at the meeting (Kanesville, 1852) to establish bylaws for the Dubuque Emigrating Co. Father of William Sutton Cooke. Husband of Sarah Cooke.
Cooke, William Sutton	Age 25. Husband of Lucy, traveled together with Dubuque Emigrating Co.
Coriell, Augustus	Went via the Isthmus. Killed trying to extinguish a lighted cigar that had landed in a powder keg. Exploded. Had made $2,400.
Coriell, Edward	Age 28. Went overland to California in 1849, returned in 1853. Lead mining experience.
Coriell, Edwin	
Coriell, J.	
Coriell, W.W.	Age 42. Newspaper editor of the *Iowa News* and the *Democratic Telegraph*. Government land surveyor, land commissioner, lawyer. Alderman in 1841.
Corkery, John	Clerk— Office of Receiver of Public Money in Dubuque. Died in California in 1868.
Courtney	Went with Great Dubuque Co. Remained in California for two years. Living in Vernon Township, Dubuque County, in 1880.
Cox, A.W.	
Cox, Joseph	Went via the isthmus.
Cox, Richard	Landowner/speculator. Went via the isthmus.
Cox, Sam	
Craig, Alexander	Age 32. The name A. Craig, Dubuque, Iowa, is carved on Register Cliff, Wyoming, dated May 19, 1850. He also signed at Independence Rock on May 26 and may also have signed at Old Castle on May 28. Died in the Civil War.
Crane, Ambrose	Doctor. Had a decided inclination for surgery.
Crane, Thomas	Experience in mining operations.
Cravier, M.	Could this be Crarier or Crevier?

Crevier, Joseph	Age 31. Experience in mining operations. Farmer from Richardsville; led a party from there to California. Returned in 1851. Found little gold.
Crozier, Mr.	Resigned as sheriff of Dubuque County on May 13, 1850, to go to California.
Curtis, H.K.	
Davis, B.G.F.	
Delay, Dennis	
Delong, H.	
Depui, James	Described as young. Father was a preacher in Dubuque. Killed by Native Americans. Left Kanesville in May 1849 with the Wisconsin and Iowa Union Co.
DeSavie, F.	Could be DeSavio?
DeSene, T.	
Desevere, T.	Experience in mining operations.
Desmoineux, Charles	Age 32. Baker and confectioner. To California in 1849 with the San Francisco Co. out of Jackson Co., Iowa.
Dexter, F.A.	Age 24. Went to California in 1851. Returned in 1861 and engaged in mining.
Digman, Michael	
Donallan	Donallan names inscribed on Independence Rock.
Donallan	
Donnelly	Friend of George Madeira
Dorsey, C.W.	Age 34. Merchant. Dubuque Emigrating Co., 1852.
Dougherty, E.C.	Town trustee at Dubuque in 1840. In California in March 1852, where he was a justice of the peace.
Douglass, Joseph M.	Went via the isthmus.
Dreibelbis, Martin	Age 42. Was a carpenter in 1850. In Dubuque in 1854 census. Left Kanesville in May 1849 with the Wisconsin and Iowa Union Co.

Drummond	Falsely reported to have been killed on the trail.
Drummond	
Dupui, James	See Depui
Eggleston	Eggleston names inscribed on Independence Rock.
Eggleston, George	Age 32. Carpenter. Dubuque Emigrating Co., 1852.
Eighmey, C.H.	Age 17. Went to California in 1851. Remained five years. Entered legal profession.
Ellis, B.F.	
Evans, J.P.	Took him eight months of the "most severe peril and hardship" on the overland route to get to Sacramento. Was Dubuque County marshal from 1856–57.
Everingham, William	Farmer. Left wife and four children to pan gold. Was in California in May 1851.
Fancher	
Fanning, James	Mayor in 1843. Alderman, prominent in business and public affairs. Dubuque County supervisor in 1837–38. Experienced in lead-mining operations.
Fearn, Jack	
Fellows, Moses	
Forbes, C.C.	
Fulweiler, Abraham	Father of John M. Fulweiler.
Fulweiler, John M.	Age 17. Went to California in 1850. More than ordinarily successful. Came back for his family and returned to California in 1852.
Gab, Doctor	
Garner, Conrad	Experience in mining operations.
Garner, J.	Experience in mining operations.
Gartrell, H.	In California in March 1852.

Joseph Gehrig, Dubuque forty-niner. *Courtesy of the Center for Dubuque History, Loras College.*

Gehrig, Joseph	Age 29. Worked in stone quarry. Returned, 1851. Bought lots for hotel. Ran Jefferson House at 7th & White Streets. During excavation for his hotel, the body of Patrick O'Connor was found.
Gillespie, Ed F.	Merchant. Drugs, medicines, books. Also "segars" and snuff.
Gilliam, William S.	Age 30. Merchant. Went via the Isthmus in 1850. He was "disgusted" with the experience
Glenat, Valentine	Age 45. Died of cholera on the trail, July 7, 1849. Merchant, dry goods, groceries. Probate judge for Dubuque Co. Experience in mining operations.
Gooddaire, Henry	
Goodrich, J.R.	
Gratiot, Charles H.	Age 35. Smelting business.
Gratiot, Henry	Age 26. Miner.
Gratney, Joseph P.	Died on the trail, July 7, 1849.
Hadley, Henry G.	Age 35. Confectioner, dry goods, groceries in Elkader, Iowa. No success in 1849, returned in 1850 and then went to Oregon in 1851.
Hammonds, S.M.	Went via the isthmus in January 1849. Back in Dubuque one year later. His experience was "anything but flattering."

Harris, Joseph	
Hart	The Harts inscribed their names on Independence Rock.
Hart, R.W.	Dealer in dry goods and clothing. Died from inflammation of the bowels six days after arriving in California.
Harvey, Mr.	
Haslett, Mr.	Died—cause undetermined.
Haughey, P.	Name carved on Register Cliff, Wyoming, May 19, 1850.
Hawley, Dean	
Heacock, Josiah (aka Joseph)	Suffered much from scurvy.
Heecock, James	
Hempstead, Joseph L.	Age 39. Went via the isthmus. Died of stomach cancer at San Francisco in April 1850.
Hewitt, Charles C.	Wagon maker? Worked to suppress cholera in Dubuque. Chaired the committee to establish governance of the Dubuque Emigrating Association. Candidate in Dubuque in 1853 for receiver of public moneys.
Higgins, Warren	Age 49. To California in 1849 and was unusually successful. Died in Dubuque Co. in 1857.
Hogan family	Louise Hogan born in 1845. To California in 1852. Later married Joel Parmeter.
Holbrook, John C.	Went to California in 1858. Arrived there March 17, 1859, where he continued as a preacher.
Hulbert	Hulbert names inscribed on Independence Rock.
Hulbert	
Hunn, J.J.	
Jalliger, Pat	To California on April 2, 1850.
Jenks, O.J.	

Jones, Augustus, Jr.	
Jones, David	
Jordan, Banion (Bunyan)	Age 39.
Jordan, George W.	Farmer. Age 37. Died May 1, 1850, of "congestion of the brain." Buried near Platte River.
Jordan, J.B.	Worked in mining, employed 8 people.
Kane, Colonel	
Karrick, Joseph	Age 26. Son of George O. Karrick. Came home with $6,000. Came back to Dubuque in a dazzling costume bespangled with gold coins. Supposed to have been at Custer's Last Stand, serving under Major Reno, and surviving.
Kennedy, T.K.	He "signed in" at Register Cliff, Wyoming, on May 19, 1850. Carved in next to him are the names of A. Craig and P. Haughey from Dubuque.
Kibbee, James.	Age 27. Died at San Diego, congestive fever.
Kirkpatrick, J.M.	
Koch, Adam	Age 19. To California in 1849.
Koozer, B.P.	In the army in California at time of the gold discovery. Previously had been a journeyman printer for the *Dubuque Tribune.*
Lane, Jim	To California in 1849 with the San Francisco Co. out of Jackson Co., Iowa.
Langdon, J.H.	
Langton, James A.	Age 27. Experience in mining operations. To California in 1849, back in Dubuque in 1851. Operated a mercantile house in Dubuque. Elected as Dubuque's town clerk in 1853.
Langworthy, Stephen C.	Age 25. Went to California via the isthmus. Left Dubuque on November 16, 1849. Engaged in mercantile pursuits in California. Accumulated a fortune. Son of Dr. Stephen Langworthy and his second wife, Jane Moreing—thus a stepbrother to the more well-known James, Lucius, Solon and Edward.
Larkin, John	

James A. Langton. *Courtesy of the Center for Dubuque History, Loras College.*

Philip C. Morheiser, Dubuque forty-niner. *Courtesy of the Center for Dubuque History, Loras College.*

Leathers, Allen	Was back in Dubuque in 1855 running the omnibus on Main St.
Livermore	
Logan, Augustus R.	Reported back in Dubuque in May 1852.
Logan, Charles	Went via the isthmus.
Long, George	
Lorimier, Charles	Age 19. Suffered from scurvy. Experience in mining operations.
De Lorimier, W.K.	Age 30. Engaged in lead smelting. To California in 1851 and back in 1853.
Madeira, Daniel	Age 16. To California in 1852.
Madeira, Francis A.	Age 21. Became a trained surveyor. To California in 1852.
Madeira, George	Age 15. To California in 1852. Established the first astronomical observatory in California.
Madeira, George Asbury, Col.	Age 51. Attorney at law. Dubuque Emigrating Co., 1852. Also had a farm, Buena Vista, south of Dubuque that he sold to go to California. Madeiras' names inscribed on Independence Rock.
Madeira, Susan	Age 41. To California in 1852.

Mahony, John	
Marshall, William	
Martin, T.O.	
Maxwell	Made about $3,000 by January 1850.
McClane, Frank	
McCoy, J.	Went with Merritt via the Isthmus on January 1, 1849.
McCraney, J.	
McCraney, Thomas	Smelting in 1833. Operated a horse ferry on the Mississippi River from Eagle Point to Sinipee.
McCraney, W.	Experience in mining operations.
McDaniel, George	From the mines.
McDaniel, Richard	
McGarvine	McGarvine's Co. from Dubuque made up of six persons.
McNair, Robert	Age 29. Miner. Dubuque Emigrating Co., 1852. Dubuquer? Or did he join at Kanesville?
McQuillan, Thomas	Age 17. Farmer. Went to California in 1852.
Merritt, William H.	Age 30. Lawyer. Sometime proprietor of *Miners' Express.* Went in January 1849 via the isthmus, where he encountered a lengthy delay. In November 1849, he had made $700. Back in Dubuque in March 1851.
Miller, Charles	Making $12 a day in California.
Miller, W.W.	
Mitchell, James	Age 34. Miner. Left Kanesville in May 1849 with the Wisconsin and Iowa Union Co.
Mobley, Edward	Went via the isthmus, where he encountered a lengthy delay. On his way home January 1850.
Molony, John	Went via the isthmus.
Morhiser, Philip C.	Age 37. Chief of police. Involved in mining. Elected as Dubuque City marshal in 1853. Dubuque County marshal from 1867–68.
Morrison, C.	

Morrison, David M.	Age 27. City marshal and collector. Dubuque Emigrating Co., 1852. Headed up Independent Dubuque Co. No. 1 when it split off from Dubuque Emigrating Co.
Morrison, E.	"From the mines."
Morrison, Guy	Age 32. Merchant. Postmaster of Dubuque. Treasurer of Dubuque Co. Reputed to be a friend of the Sauk Indian chief Black Hawk.
Morrison, Jesse	Morrison names inscribed on Independence Rock. "From the mines."
Morrison, John	"From the mines."
Morrison, Mary Elmira	Age 26. Wife of Guy Morrison. Arrived in California in 1854 via the Isthmus of Nicaragua.
Morse, A.C.	Author of *Journal of an Overland Trip to California*, as reported in a series of installments in the *Miners' Express* in 1850. Was a reporter for the *San Francisco Daily Herald* in 1851.
Moyea, George	
Myers	Myers names inscribed on Independence Rock.
Myers	
Newcomer, Thomas	Died of cholera. Time of illness was 30 hours.
Newton	In California in March 1852.
Nichol, James	
Nowlin, Bennett	Captain Nowlin was a member of the committee to lead the Dubuque Emigrating Association to California.
O'Brien, Edwina	Daughter of Dr. J. O'Brien
O'Brien, James	Doctor. Took his family to California in 1849. Daughter named Edwina.
O'Connell, J.	
O'Ferrall, F.K.	Mayor of Dubuque, 1844–46. Bought and shipped lead. Engaged in real estate and merchandising. One of Dubuque's most prosperous merchants.

O'Ferrall, F.K., Jr.	One of the F.K. O'Ferralls reportedly died—which one is unknown. Died December 5, 1851, at age 41.
O'Moore, Rory	
Parker, John	Captain Parker was a member of the committee to lead the Dubuque Emigrating Association to California.
Paul, W T.	
Peckham, G.W.	
Peckman, Mr.	Went via the isthmus.
Perdreauville, Leon	
Perrin	To California in 1852. Perrin names inscribed on Independence Rock.
Perrin, Aaron	Married Lucy A. McMahon in Dubuque in 1847. Traveled with Dubuque Emigrating Co. in 1852. Headed up Independent Dubuque Co. No. 2 when it split off from the Dubuque Emigrating Co.
Philips, Isaac	
Plum, N.D.	
Plumbe, John	Age 40. Later a photographer scouting topography for a railroad route. Elected city surveyor in Sacramento. Returned to Dubuque in 1854.
Porter, H.	
Quigley, Daniel	Age 19.
Quigley, John P.	Age 25. Went in 1850, came back to Dubuque in 1855. Became a doctor. Son of early pioneer and prominent Dubuquer Patrick Quigley. City alderman for seven years.
Ray, Mr.	Died. Killed by Native Americans.
Read, Joseph Robert	Married in Dubuque in 1845 before moving to California in 1849 or 1850.
Reno, L.W.	
Reynolds, R., Jr.	
Reynolds, R., Sr.	

Reynolds, Richard	Not one of the R. Reynolds listed above.
Rickard, Antoine	
Rickets, Isaac	
Rickey, James.	Age 50. Older brother of Thomas Brinley Rickey. James sold his tavern in Dubuque and left with his wife and three children. Split off near Salt Lake and went to Oregon.
Rickey, James Allen	Age 17. The son of Thomas Brinley Rickey. Dubuque Emigrating Co., 1852. Joined up at Kanesville.
Rickey, James Madison	Age 12. The son of James Rickey.
Rickey, Thomas Brimley	Age 45. Miner. Dubuque Emigrating Co., 1852. Joined up at Kanesville. Both Rickeys' names were inscribed on Independence Rock. Made the second trip with his pregnant wife and 10 children.
Roberts, Fred	Selling stoves in May 1848.
Roberts, George	
Roberts, William	The Robertses inscribed their names on Independence Rock.
Robison, Warren Only	Age 30 when he went to California in 1852. Was engaged in lead mining for two years in Dubuque.
Saucier, Henry	
Seere, J.B.	Experience in mining operations.
Shervin, P.	Experience in mining operations.
Shields, F.M.	Doctor. To California in 1852.
Shinn, Asa	Age 28. Married Azariah Morgan in Dubuque in 1849. Dubuque Emigrating Co., 1852.
Shipton, John	Led a company of five persons.
Simpare	
Sims, James	Age 24. To California in 1850, back in Dubuque in 1855. Mill operator at Catfish Mill in Rockdale, Thompson's Mill at Sageville and then farming.

Singleton, Michael	
Sloan, Henry	
Smith, John	Engaged in lead mining. To California in 1849, where he was unusually successful.
Spensley, James	Age 16. His oxen died in Nebraska, and he walked the rest of the way to California.
Starr, John M.	Dubuque Emigrating Co., 1852. Musician in Civil War, 46th Iowa Infantry.
Stewart, Moses	
Stiger	
Stough, William	
Strohl	
Strong, George	
Sullivan, John	
Summers, Peter	Age 45. Farmer. Cascade Township. To California in 1852, back in 1855. He "bettered himself."
Sutherland	
Thomas, Edward	
Treanor, Hugh	Age 47. Engaged in mining and the grocery trade. City alderman for eight years.
Turner	
Upchurch, William L.	Age 54. Married Elizabeth Smith in Dubuque in 1848. Dubuque Emigrating Co., 1852.
Van Hagan, Charles C.	Age 19. Dubuque Emigrating Co., 1852.
Van Hagan, Isaac. N.	Age 24. Dubuque Emigrating Co., 1852. Merchant.
Van Hagan, J.B.	City assessor in 1851. Dubuque Emigrating Co., 1852. Van Hagan names are inscribed on Independence Rock.
Wallace, Pat	
Waller, Robert	
Walsh, John	Died on the prairie of unknown causes.
Walter, Mr.	
Webb	Webb's Co. from Dubuque.

Wharton, John	Suffered from scurvy.
Wheeler, Loring	Age 50. Chief justice of Dubuque County, served in the first legislative assembly of Iowa from Dubuque Co. Trustee of the town of Dubuque in 1839. Captain of the security force at the execution of Patrick O'Connor.
White, Volney	Age 17.
Whitesides, Abraham R.	Age 46. Experience in mining operations. One of the Whitesides (which one not specified) was presumed killed by Native Americans. Was at the Dubuque Mines in 1830.
Whitesides, E.M.	Experience in mining operations. Was at the Dubuque Mines in 1830.
Wight, Calmes L.	Age 24. Captain in Mexican War. On his way to California in April 1849. Died before reaching St. Louis.
Wilde, George	Age 28. Engaged in smelting. Upon his return from California to Dubuque, he was then in mercantile and later in the livery business. Alderman in 3rd Ward.
Wilkinson, Thomas	
Willer, C.	
Williams S.G.	By March 1850, he had made $2,000 keeping bar on a boat to Stockton.
Woodward	Confectioner.
Woodward	The Woodwards inscribed their names on Independence Rock.
Yates, John	Erected a new dwelling house in Dubuque in 1850. Dubuque Emigrating Co., 1852.
Yates, Robert	Age ca. 50. To California in 1852. Died there in 1854.
Zeb (no last name given)	
Z., L.	Had been employed by the *Dubuque Tribune*. To California in 1852.

NOTES

PREFACE

1. Two sources provide differing numbers for the population for 1850; one indicates 3,108, the other 4,071. This study will use the 4,000 figure, as it comports more closely with the continuing growth of the city and the county as seen in the years before and after 1850. Thus, by way of estimates from various sources, we have Dubuque city population for the years 1849 to 1852 as 3,500, 4,071, 6,000 and 7,000, respectively.
2. William Bothwell to Dr. Horr, September 10, 1849, Feather River, California, *Dubuque Tribune*, March 1, 1850, p. 2, cols. 5–6; James Langton to Den, *Miners' Express*, December 12, 1849, March 6, 1850, p. 1, col. 4; Henry Gooddaire to Basil Gooddaire, February 16, 1850, Weber Creek, California, *Miners' Express*, May 15, 1850, p. 1, cols. 3–4.

CHAPTER 1

3. More detailed treatments of Julien Dubuque can be found in *Julien Dubuque* by Thomas Auge and *Julien Dubuque, His Life and Times* by Richard Herrmann.
4. U.S., *Public Statutes at Large*, 7, 374–76.
5. Contrary to what earlier writers assert—that official settlement of these lands was not to occur that soon—there is sufficient evidence to show that settlement by whites was not only sanctioned but also encouraged by the government. See Childs, *Dubuque*, 15, note 10.
6. Conzett, *Recollections of People and Events*, 67. See throughout this work for what amounts to a "City Directory" for the period of time prior to

the first published *Dubuque City Directory* in 1856. One must be cautious, however, as the chronology of his descriptions jumps around.

7. *History of Dubuque County*, 527–28. But compare this to his much more optimistic statement on p. 529 of the same work.
8. Johnson, *Warriors into Workers*, 41, n. 36. The 1860 census identified 389 people occupied as miners in Dubuque. George Ord Karrick employed 41 men at his Julien Mining Co. There were 30 miners identified as businessmen who owned property worth more than $3,000.
9. Heyl, *Geology of the Upper Mississippi Valley*, 74.
10. *Dubuque Tribune*, September 29, 1848, p. 2, col. 3.
11. Childs, *Dubuque*, 117, 141.
12. Mahoney, *River Towns*, 223.
13. *Memoranda. &c. Journey from Baltim*[or]*e to St. Paul's Minnesota.* May 7 to June 20, 1851. Manuscript in possession of Minnesota Historical Society. As quoted in Petersen, *Steamboating*, 340.
14. Auge, *Comparison of 1850 and 1860 Census.*
15. For city data, see *Miners' Express*, October 3, 1848, p. 3, col. 5, for 1848; Oldt, *History of Dubuque County*, for 1849; *Census 1880* and Childs, *Dubuque*, 94 for 1850; and Childs, *Dubuque*, 108 for 1851 and 1852. For county data, see Iowa, *Census Returns, 1859*, table facing p. 3. For state data, see Iowa. *Census of Iowa for 1880*, 168 and 474.
16. DeBow, *Statistical View*, 231.
17. See Adelmann, *Dubuque Shot Tower*, for an extensive investigation of this Dubuque landmark.
18. Holbrook, *Recollections of a Nonagenarian*, 64.
19. Conzett, *Recollections of People and Events*, 68.
20. *Miners' Express*, February 23, 1848, p. 2, col. 4.
21. As cited in Abbot, *Boosters and Businessmen*, 189.
22. *Miners' Express*, December 19, 1849, p. 2, col. 1. A similar report for 1850 can be found in the *Dubuque Tribune*, November 13, 1850 p. 2, cols. 1–2. Broader distribution of the *Miners' Express* report appeared in the *Western Journal* 3, no. 5 (February 1850): 326–27.

CHAPTER 2

23. U.S., *Statutes at Large*, 9, 922–42. It is interesting to note here what is probably a little-known fact. By this treaty, the Rio Grande became the southern border of Texas, and the United States acquired the present

states of California, Arizona, New Mexico, Nevada and Utah for a total of 1.2 million square miles. Compare this to the more famous acquisition of the Louisiana Territory, with a total of only 828,000 square miles.

24. Holliday, *Rush for Riches*, 59. New Helvetia was so named because John Sutter was from Switzerland—*Helvetia* being the Latin word for part of present-day Switzerland and lands adjacent to it.
25. *California Star*, "Great Express Extra," as quoted in Holliday, *Rush for Riches*, 60.
26. Polk, *Message*, 56–64.
27. *Miners' Express*, December 26, 1848, p. 2, col. 6.

CHAPTER 3

28. Bonson, *Diary*, December 29, 1848.
29. *Miners' Express*, April 24, 1849, p. 2, col. 4.
30. Bonson, *Diary*, April 10, 1849, and February 14, 1850. No record has been found of Delaney going to "Callifronia."
31. Crétin to his sister, April 12, 1849, as quoted in Hoffmann, *Church Founders*, 248; Perrodin to Loras, February 6, 1849, Otting, *Letters to a Pioneer Bishop*, 665; and Trevis to Loras, September 10, 1851, Otting, *Letters to a Pioneer Bishop*, 821. Likewise, Father J. McEvoy proposed a scheme to Bishop Loras in which he would go to California and collect enough gold to pay for St. Patrick's Church and a new cathedral in Dubuque. The bishop was not persuaded. Hoffmann, *Centennial History*, 89.
32. Copeland, "The Cornish in Southwest Wisconsin," 330.
33. Langworthy, *Diary*, September 16, 1860. As cited in Mahoney, *Provincial Lives*, 115.
34. Manly, *Death Valley in '49*, 58.
35. Bonson, *Diary*, May 19, 1849; June 25, 1849; July 3, 1849; July 4, 1849. Spelling corrections are not made in quoting Bonson in order to provide "flavor" to the man and the times.
36. See Scharnau, "From Pioneer Days," for an excellent investigation of the laborer in Dubuque at mid-century. For the relative purchasing power of money in the United States, see Historical Currency Conversions at futureboy.homeip.net/fsp/dollar.fsp?quantity=1&currency=dollars&from.
37. Lorch, "Iowa and the California Gold Rush," 367. As found in the [Andrew, Iowa] *Western Democrat* of March 1, 1850. See also Lorch, 317 and 343, and Taylor, *From Lead Mines to Gold Fields*, 30.

38. McKinstry, *California Gold Rush*, 68.
39. Coon, *Journal of a Journey*, 180.
40. Allen to Weatherby, December 13, 1849. *Miners' Express*, March 13, 1850, p. 2, col. 6.

Chapter 4

41. Bothwell to Dr. Horr, September 10, 1849, *Dubuque Tribune*, March 1, 1850, p. 2, cols. 5–6. By 1855, Bothwell was back in Dubuque and was ruling elder of the Second Presbyterian Church.
42. Negotiating the Rocky Mountains was not as difficult as one might imagine, using the preferred path through South Pass in southwestern Wyoming. Heacock, *To California by Covered Wagon*, July 15, 1849, describes it as "about 20 miles wide and so smooth and level that a person scarcely knows when they are on the extreme summit."
43. Holliday, *Rush for Riches*, 103.
44. Heacock, *To California by Covered Wagon*, June 25, 1849.
45. *Miners' Express*, June 13, 1849, p. 1, col. 6. Reprinted from the St. Joseph's *Gazette.*
46. For an extensive descriptive narrative of what these forty-niners faced at Chagres, see *Dubuque Tribune*, January 12, 1849, p. 2, col. 4.
47. Gillespie to Benton, May 18, 1849, *Miners' Express*, June 20, 1849. p. 2.
48. *Miners' Express*, February 6, 1849, p. 1, col. 7.
49. Unruh, *Plains Across*, 403, provides a table showing the number of travel days estimated for the various segments of the California Trail.
50. For a detailed study on the published information available to the forty-niners at that time, see Stillson, *Spreading the Word.*
51. *Dubuque Tribune*, January 26, 1849, advertises two books available: p. 2, col. 3, Edwin Bryant, *What I Saw in California*; and p. 3, col. 1, E. Seymour, *Emigrant's Guide to the Gold Mines of California* "to be published in a few days. Buy from Seymour at Galena, Ill. for 30 cents." See also *Dubuque Tribune*, February 2, 1849, p. 3, col. 2, D'Alvear, *Gold seeker's Guide.* $1.00. and a *"Goldometer."* And p. 2, col. 2, *Fremont's Map of Upper California and Oregon*, as well as Joseph E. Ware, *The Emigrants' Guide to California, containing every point of information for the Emigrant Including Routes, Distances, Water, Grass...with a Large Map of Routes, and Profiles of Country, &c—with Full Directions for Testing and Assaying Gold and Other Ores.*
52. Wagner and Camp, *Plains & the Rockies*. See especially 307–63 for the years 1848–49. See also Kurutz, *California Gold Rush*, for an extensive

listing of books and pamphlets on the gold rush—707 titles—though some of these were published in later years.

53. For an excellent survey of guides available in 1848 and 1849, with an emphasis on the maps contained therein, see Wheat, *Mapping the Trans-Mississippi West*, 3, 49–109.

54. Levy, "Intrepid Females of Forty-Nine," 151–52.

55. West, "Plagues of the Gold Rush," 10.

56. These three had left camp to retrieve stolen cattle and were never heard from again. They were presumed killed by the "root diggers," a term applied generically to various Native American tribes in the central California region.

57. Shoemaker, *Overland Trail Diary*. "Here is the grave of George Washington Gordon from de buke Co. io.wa. He died may 1st 18.50 with the congestion on the brain." Diary entry for May 11, 1850, 47. Shoemaker was then 302½ miles from the Missouri River.

58. *Miners' Express*, May 8, 1850, p. 1, col. 4.

59. *Bylaws and Resolutions*. Also to be found in Cooke, *Crossing the Plains*, 18–19. The compact of another company with Dubuque members was the Wisconsin and Iowa Union Company. Their agreement was printed in the Kanesville *Frontier Guardian*, May 30, 1849, p. 3, col. 4.

Chapter 5

60. Levy, "Intrepid Females of Forty-Nine," 151–61.

61. Merritt to [addressee not given], January 12, 1850. *Miners' Express*, April 17, 1850, p. 2, col. 6.

62. West, "Plagues of the Gold Rush," 10. The letter of A.C. Morse of December 12, 1850, to Dr. Holt reports on this cholera disaster in Sacramento. *Miners' Express*, February 12, 1851, p. 1, cols. 4–5. See also Roth, "Cholera, Community, and Public Health," who addresses specifically this San Francisco/Sacramento event.

63. Hempstead to McKnight, October 15, 1849. *Miners' Express*, January 2, 1850, p. 1, col. 3.

64. Kurutz, *California Gold Rush*. See Holliday, *Rush for Riches*, "Introduction," xxvi, n. 2.

Chapter 6

65. *Miners' Express*, April 17, 1849, p. 2, cols. 4–6.
66. The country of Panama is about 425 miles long and, on average, about 70 miles wide. It lies generally in an east–west direction. Chagres was the coastal village on the Atlantic Ocean at which the forty-niners disembarked to begin their journey across the Isthmus of Panama. It was small (a population of about four hundred), backward, devoid of facilities and totally unprepared to deal with the large influx of gold seekers. To look at the bright side, most were so eager to get to California that they quickly left Chagres to cross the isthmus to find transportation up the Pacific coast to San Francisco. Panama City (simply called Panama by Merritt and other forty-niners) is on the Pacific coast and lies south and east of Chagres—not to the west, as one might think.
67. *Miners' Express*, May 1, 1849, p. 2, cols. 4–7.
68. The Republic of New Granada consisting mainly of today's Colombia and Panama, along with parts of Ecuador and Venezuela, was created in 1830 and lasted until 1858. Panama remained for a time part of Colombia and later achieved independence on November 3, 1903. The Panama Canal was built by the United States from 1904 to 1914. It was operated by the United States until the canal and the Canal Zone in which it operated were returned to Panama on December 31, 1999.
69. Panama City was the site on the Pacific side of the Isthmus of Panama that became the western terminus for the isthmian crossing—the companion to Chagres on the Atlantic side. It was here that the earliest forty-niners expected to find vessels to transport them to California. In many, many cases, their expectations were unmet, there yet being few sailing or steamboats operating on this route, and almost no regularity in their scheduling. The result was that many forty-niners were stranded on the isthmus for weeks and even months before getting off. For an eyewitness description of Panama, see the letter of Edward Mobley, March 20, 1849. *Dubuque Tribune*, May 25, 1849, p. 2, cols. 1–2.
70. A small village at the western end of Lake Gatun. The isthmian crossing from Gorgona to Panama City had to be completed on overland trails.
71. *Miners' Express*, May 8, 1849. p. 1, cols. 4–5.
72. Ibid., June 20, 1849, p. 1, col. 5.
73. Ibid., p. 1, col. 6.
74. Ibid., June 13, 1849, p. 1, col. 5.
75. Ibid., October 10, 1849, p. 3, col. 1.

76. Ibid., January 30, 1850, p. 2, col. 5.
77. Ibid., February 6, 1850, p. 2, col. 5.
78. Ibid., April 17, 1850, p. 2, col. 6.

CHAPTER 7

79. *Dubuque Tribune*, April 7, 1852, p. 1, col. 5.
80. Ibid., May 12, 1852, p. 2, col. 7.
81. Ibid., August 4, 1852, p. 1, cols. 4–5.
82. Rock Avenue was a well-known site on the California Trail as the forty-niners passed through what is now Natrona County in Wyoming. It was a portion of the trail where rocks rose twenty to forty feet above the surrounding terrain on either side of a pathway about one hundred feet wide.
83. *Dubuque Tribune*, June 17, 1852, p. 1, col. 2.
84. Devil's Gate and Independence Rock, located within about five miles of each other, were two of the most famous landmarks on the Overland Trail in Wyoming some fifty miles southwest of Casper. Independence Rock was the first to be encountered. It is a large granite mound of a rock rising some 125 feet above the ground. Many of the immigrants carved their names in the rock, some of which can still be seen. Its name may have come from the forty-niners who wanted to be at this point on the trail by July 4, Independence Day, in order to pass through the Sierra Nevada Mountains prior to the snowfalls. Devil's Gate was found about five miles farther on. It was a sharp, narrow, 370-foot-high cut through a mountain created by the Sweetwater River running through it. It was about 30 feet wide at the bottom, opening to about 300 feet at the top. The trail did not take the pioneers through Devil's Gate.
85. *Dubuque Tribune*, August 18, 1852, p. 1, col. 2.
86. Ibid., September 16, 1852, p. 1, cols. 2–4.
87. Ibid., January 19, 1853, p. 2, col. 4.

CHAPTER 8

88. Sources for the letters are Keller, *Anna Morrison Reed*, and Hartman, *Preston's*.
89. Weston is located about halfway between Independence and St. Joseph, Missouri.

90. Keller, *Anna Morrison Reed*, 199–201.
91. Dixon Ravine lies about seventy miles due north of Stockton.
92. Hartman, *Preston's*, 90.
93. San Juan del Norte on the Caribbean coast of Nicaragua. Thus, she did not cross via the Isthmus of Panama but, rather, across Nicaragua.
94. As the crow flies, Bidwell's Bar was about six miles from Dixon Ravine, where Mary's husband, Guy, was working for gold.
95. Hartman, *Preston's*, 53–55.
96. Ibid., 59–60.
97. Ibid., 92.
98. Ibid., 61–62.

CHAPTER 9

99. Johnson, *Warriors into Workers*, 41, n. 36. The 1860 census identified 389 people occupied as miners in Dubuque. George Ord Karrick employed 41 men at his Julien Mining Co. There were 30 miners identified as businessmen who owned property worth more than $3,000.
100. *Dubuque Tribune*, February 9, 1849, p. 2, col. 5. See also *Andrew Western Democrat and Common School Journal*, February 22, 1850, p. 2, col. 1.
101. There are nearly three hundred forty-niners identified by name in this study. It is likely that some of these were not from Dubuque itself but from nearby areas and villages that joined the Dubuque companies. Undoubtedly, there were more whose names do not appear in the historical record.
102. There were many who crossed at Dubuque who left no written record of doing so, and others who left a written record unknown to historical researchers. Mattes records a number of diarists who made mention of Dubuque on their journey. See entry numbers 660, 894, 1152, 1153, 1169, 1180, 1255, 1276, 1292, 1327, 1375, 1428 and 1450.
103. Brick, *Iowa Underground*, 107. Further statements regarding the loss of Iowa miners to the gold rush are found in Cole, *History of the People of Iowa*, 238; Hall and Whitney, *Report of the Geological Survey of the State of Iowa*, vol. 1, 466–67; and Keyes, "History of Geographic Development," 107. Other comments regarding the decline of mining for Dubuque are Aitchison, "Geographic Factors," 24; Loras, *Foundations*, 690; and *Dubuque Folklore*, 94.
104. *Miners' Express*, October 23, 1850, p. 2, col. 4. Contrast this with the much more negative report of *History of Dubuque County*, 527–28, and his more optimistic statement on p. 529.

105. *History of Dubuque County*, 529.
106. *Miners' Express*, April 24, 1852, p. 2, col. 1.

Chapter 10

107. For an opposing viewpoint, see Childs, *Dubuque*, 87, where he states, "Many of them returned in a year or two with a golden reward" and further, "Several of our prominent businessmen owe the commencement of their fortunes to their successful adventures in California." Unfortunately, he does not give a dollar amount for the golden rewards, nor does he name anyone.

Chapter 11

108. *History of Dubuque County*, 527–28.
109. The analysis of characteristics regarding the gold seekers from Dubuque was made based on the data available. For example, of the total of 298 identified forty-niners, the age of only 81 has been uncovered. While occupations or professions have been identified for some, there may well be more who were so engaged, but the research has not been able to make such a connection. Therefore, the analysis can be taken as only indicative.

Bibliography

Abbott, Carl. *Boosters and Businessmen: Popular Economic Thought and Urban Growth in the Antebellum Middle West*. Westport, CT: Greenwood Press, 1981.

Adelmann, John, ed. *The Dubuque Shot Tower.* Charleston, SC: The History Press, 2011.

Aitchison, Alison Esther. "Geographic Factors in the History of Dubuque County, Iowa." Master's thesis, University of Chicago, 1914.

Andrew Western Democrat and Common School Journal. Andrew, IA.

Atwater, Caleb. *Writings of Caleb Atwater.* Columbus, OH: self-published, 1833.

Auge, Thomas E. "Comparison of 1850 and 1860 Census." Unpublished typescript. Dubuque, IA: Center for Dubuque History, Loras College, n.d.

Auge, Thomas E., Michael D. Gibson and Robert F. Klein. *The Mines of Spain in Its Historical Context.* Final Report Submitted to the Conservation Commission, Department of Natural Resources, State of Iowa. Dubuque, IA: Center for Dubuque History, Loras College, 1986.

Belcher, Wyatt. *Economic Rivalry Between St. Louis and Chicago, 1850–1880.* (Studies in History, Economics and Public Law, no. 529). New York: Columbia University Press, 1947.

Billington, Ray Allen. "Books That Won the West: The Guidebooks of the 49ers and 59ers." *American West* 4 (August 1967): 25–32, 72–75.

Blair, Roger P. "The Doctor Gets Some Practice; Cholera and Medicine on the Overland Trails." *Journal of the West* 34, no. 1 (January 1997): 54–66.

Bonson, Richard. "Diary" [unpublished typescript], 1840–1904. DVD available at Center for Dubuque History, Loras College Library.

Booth, Edmund. *Edmund Booth (1810–1905) Forty-niner: The Life Story of a Deaf Pioneer.* Stockton, CA: San Joaquin Pioneer and Historical Society, 1953.

Brick, Greg A. *Iowa Underground: A Guide to the State's Subterranean Treasures.* Black Earth, WI: Trails Books, 2004.

Brown, Randy. *Historic Inscriptions on Western Emigrant Trails.* Independence, MO: Oregon-California Trails Association, 2004.

Browning, Peter. *To the Golden Shore: America Goes to California.* Lafayette, CA: Great West Books, 1995.

Bylaws and Resolutions of the Dubuque Emigrating Company to California. [1852]. In *Covered Wagon Women: Diaries & Letters from the Western Trails, 1840–1890.* Edited and compiled by Kenneth L. Holmes. Vol. 4, *1852: The California Trail*, 209–95. Glendale, CA: Arthur H. Clark, 1985.

Calvin, Samuel, and H.F. Bain. "Geology of Dubuque County." Iowa Geological Survey. *Annual Report* 10 (1899): 380–622. Des Moines: Iowa Geological Survey, 1900.

Chaffin, Tom. *Pathfinder: John Charles Frémont and the Course of American Empire.* New York: Hill and Wang, 2002.

Chatterley, L. Matthew. *Wend Your Way: A Guide to Sites Along the Iowa Mormon Trail.* Ames: Iowa State University Press, 2000.

Chetlain, Augustus L. *Recollections of Seventy Years.* Galena, IL: Gazette Publishing Co., 1899.

Childs, Chandler C. *Dubuque: Frontier River City.* Edited by Robert F. Klein. Dubuque, IA: Loras College Press, 1984.

Chouteau v. Molony. U.S. Reports. December term, 1853 (Howard 16), 203–42; 57 U.S. 203; 14 L. Ed. 905.

Cole, Cyrenus. *A History of the People of Iowa.* Cedar Rapids, IA: Torch Press, 1921.

"Commercial Statistics." *Western Journal* 3, no. 5 (February 1850): 333.

Conzett, Josiah. "Recollections of People and Events of Dubuque, Iowa, 1846–1890." [unpublished typescript]. Center for Dubuque History, Loras College.

Cooke, Lucy Rutledge. *Crossing the Plains in 1852: Narrative of a Trip from Iowa to "The Land of Gold," as Told in Letters Written during the Journey.* Fairfield, WA: Ye Galleon Press, 1988.

———. *Letters on the Way to California.* In *Covered Wagon Women: Diaries & Letters from the Western Trails, 1840–1890.* Edited and compiled by Kenneth L. Holmes. Vol. 4, *1852: The California Trail*, 209–95. Glendale, CA: Arthur H. Clark, 1985.

Coon, Polly. *Journal of a Journey over the Rocky Mountains.* In *Covered Wagon Women: Diaries & Letters from the Western Trails, 1840–1890*. Edited and compiled by Kenneth L. Holmes and David C. Duniway. Vol. 5, *1852: The Oregon Trail*, 173–206. Glendale, CA: Arthur H. Clark, 1986.

Copeland, Louis Albert. "The Cornish and the Settlement of the Lead Region of Southwest Wisconsin." Bachelor's thesis, University of Wisconsin, 1896.

———. "The Cornish in Southwest Wisconsin." Wisconsin Historical Society *Collections* 14 (1898): 310–34.

Cowperthwait, Thomas. *A New Map of the State of Iowa.* Philadelphia: Cowperthwait, 1850.

Crutchfield, James A., ed. *The Way West: True Stories of the American Frontier.* New York: Tom Doherty Associates, 2005.

Daily Alta California. [San Francisco.]

DeBow, J.D.B. *Statistical View of the United States…Being a Compendium of the Seventh Census* [1850]. Washington, D.C.: Beverley Tucker, 1854.

"The Destiny of New Orleans." *DeBow's Review* 10, no. 4 (1851): 440–45.

Dixon, J.M. *The Valley and the Shadow: Comprising the Experiences of a Blind Ex-editor, a Literary Biography, Humorous Autobiographical Sketches, a Chapter on Iowa Journalism, and Sketches of the West and Western Men.* New York: Russell Brothers, 1868.

Dubuque. Archdiocese. Archives. Crétin Correspondence and Perrodin Correspondence.

Dubuque, City. *Book of Ordinances*. Various years.

Dubuque Daily Herald.

Dubuque Folklore: A Contribution from American Trust and Savings Bank to Dubuqueland in Celebration of Our Nation's Bicentennial, 1776–1976, by Don Leopold. Dubuque, IA: American Trust and Savings Bank, 1975.

Dubuque Tribune.

Erb, Susan M. *On the Western Trails: The Overland Diaries of Washington Peck.* Edited and with biographical commentary. Norman, OK: Arthur H. Clark, 2009.

Fairchild, Lucius. *California Letters.* Edited with notes and introduction by Joseph Schafer. (Wisconsin Historical Publications, *Collections*, 31). Madison: State Historical Society of Wisconsin, 1931.

Federal Writers' Project. *Galena Guide*. Sponsored by the City of Galena. Chicago, 1937.

Fiedler, George. *Mineral Point: A History.* Mineral Point, WI: Historical Society, 1962.

Fonda, John H. "Early Wisconsin." Wisconsin Historical Society, *Collections* 5 (1907): 205–84. Madison: Published by the society, 1907.

Fremont, John Charles. *Geographical Memoir upon Upper California, in Illustration of His Map of Oregon and California.* 30th Congress, 1st session. Senate Miscellaneous Document 148. Washington, D.C.: Wendell and Van Benthuysen, 1848. Serial 511.

———. *A Report on an Exploration of the Country Lying Between the Missouri River and the Rocky Mountains, on the Line of the Kansas and Great Platte Rivers…* 27th Congress, 3rd session. Senate Document. 243. Washington, D.C.: Printed by order of the United States Senate, 1843. Serial 416.

———. *Report of the Exploring Expedition to the Rocky Mountains in the Year 1842, and to Oregon and North California in the Years 1843–1844.* 28th Congress, 2nd session. Senate Executive Document 174. Washington, D.C.: Gales & Seaton, 1845. Serial 461. Also issued commercially: New York: D. Appleton & Co., and Philadelphia, G.S. Appleton, 1846.

Frontier Guardian [Kanesville, Iowa.]

Gara, Larry. "Gold Fever in Wisconsin." *Wisconsin Magazine of History* 38, no. 2 (Winter 1954–55): 106–8.

Gates, Paul Wallace. "Land Policy and Tenancy in the Prairie States." *Journal of Economic History* 1, no. 1 (May 1941): 60–82.

Glines, I.W. "Reminiscences of Early Times in the Lead Mines of Illinois and Wisconsin." (Paper no. 10). *The Pick and Gad* [Shullsburg, WI], May 29, 1884.

Hall, James, and J.D. Whitney. *Report of the Geological Survey of the State of Iowa Embracing the Results of Investigations Made During Portions of the Years 1855, 56, & 57.* Published by Authority of the Legislature of Iowa, 1858.

Harstad, Peter T. "Disease and Sickness on the Wisconsin Frontier: Cholera." *Wisconsin Magazine of History* 43 (1960): 203–20.

Hartman, Gerda Preston, ed. *Preston's: Thy Affectionate Family, with Love. Books I–II.* Dubuque, IA: The Editor, 1987–93.

Heacock, Josiah. "To California by Covered Wagon." Diary. [Typescript of the manuscript diary account of his journey from Dubuque to California. Entries dated from April 23, 1849, to March 3, 1850]. Typescript at State Historical Society of Iowa, Iowa City.

Heckman, Marlin L. *Overland on the California Trail, 1846–1859: A Bibliography of Manuscript & Printed Travel Narratives.* Foreword by Louis L'Amour. Glendale, CA: Arthur H. Clark, 1984.

Herrmann, Richard. *Julien Dubuque, His Life and Adventures.* Dubuque, IA: Times-Journal Company, 1922.

Heyl, Allen V., et al. *The Geology of the Upper Mississippi Valley Zinc-Lead District.* U.S. Geological Survey. Professional Paper, 309. Washington, D.C.: United States Government Printing Office, 1959.

The History of Dubuque County, Containing Biographical Sketches of Citizens, War Record of Its Volunteers in the Late Rebellion, General and Local Statistics, Portraits of Early Settlers and Prominent Men, [etc.]. Chicago: Western Historical Company, 1880.

History of Grant County, Wisconsin. Chicago: Western Historical Company, 1881.

History of Jo Daviess County, Illinois. Chicago: H.F. Kett, 1878.

Hoffman, Abraham. "Spreading the News; California's Gold Messengers of 1848." In *The Way West: True Stories of the American Frontier.* Edited by James A. Crutchfield. New York: Tom Doherty Associates, 2005.

Hoffmann, M.M. *Antique Dubuque.* Dubuque, IA: Telegraph-Herald Press, 1930.

———. *Centennial History of the Archdiocese of Dubuque.* Dubuque, IA: Columbia College Press, 1938.

———. *Church Founders of the Northwest.* Milwaukee, WI: Bruce, 1937.

Holbrook, John C. *Recollections of a Nonagenarian of Life in New England, the Middle West, and New York, Including a Mission to Great Britain in Behalf of the Southern Freedmen, Together with Scenes in California.* Boston: Pilgrim Press, 1897.

Holliday, J.S. *Rush for Riches: Gold Fever and the Making of California.* Berkeley: Oakland Museum of California, University of California, 1999.

Ingalls, Walter Renton. *Lead and Zinc Mining in the U.S.: Comprising an Economic History of the Mining and Smelting of the Metals and the Conditions Which Have Affected the Development of the Industries.* New York: Hill Publishing Company, 1908.

Iowa. Census Bureau. *The Census Returns of the Different Counties of the State of Iowa for 1859.* Des Moines, IA: Teasdale, 1859.

Iowa. Census. *Census of Iowa for 1880, and the Same Compared with the Findings of Each of the Other States, and also with All Former Enumerations of the Territory Now Embraced Within the Limits of the State of Iowa, with Other Historical and Statistical Data.* John A.T. Hull, Secretary of State. Des Moines, IA: F.M. Mills, 1883.

Iowa Writers' Program of the Works Projects Administration in the State of Iowa. *Dubuque County History, Iowa.* Jessie M. Parker, State Superintendent of Public Instruction, Statewide Sponsor of the Iowa Writers' Program. Sponsored by Joseph Flynn, County Superintendent of Schools, Dubuque County. Dubuque, IA, 1942.

Irving, Washington. *Astoria or, Anecdotes of an Enterprise Beyond the Rocky Mountains*. Philadelphia: Carey, Lea & Blanchard, 1836.

———. *The Rocky Mountains, or, Scenes Incidents and Adventures in the Far West. Digested from the Journal of Captain B.L.E. Bonneville of the Army of the United States, and Illustrated from Various Other Sources.* Philadelphia: Carey, Lea & Blanchard, 1837.

———. *A Tour on the Prairies.* (The Crayon miscellany, 1). Philadelphia: Carey, Lea & Blanchard, 1835.

Jennings, Joseph C. *City of Dubuque, Iowa, 1852*. [Broadside printed map.] N.p., 1852. Reprinted in reduced format. Dubuque, IA: Center for Dubuque History, 1979.

Johnson, Russell L. *Warriors into Workers: The Civil War and the Formation of Urban-Industrial Society in a Northern City.* New York: Fordham University Press, 2003.

Keller, John E. *Anna Morrison Reed, 1849–1921*. Lafayette, CA: self-published, 1979.

Keyes, Charles. "History of Geographic Development in Iowa." Iowa Geological Survey. *Annual Report* 22 (1912): 15–155. Des Moines, published for the Iowa Geological Survey, 1913.

Kurutz, Gary F. *The California Gold Rush: A Descriptive Bibliography of Books and Pamphlets Covering the Years 1848–1853*. San Francisco: Book Club of California, 1997.

Langworthy, Solon. "Diary." [unpublished manuscript.] Iowa City: State Historical Society of Iowa, n.d.

Lead Region Historic Trust. Shullsburg, WI. Website. www.leadregiontrust.org.

Levy, JoAnn. "The Intrepid Females of Forty-Nine." In *The Way West: True Stories of the American Frontier.* Edited by James A. Crutchfield, 151–69. New York: Tom Doherty Associates, 2005.

———. *They Saw the Elephant: Women in the California Gold Rush.* Hamden, CT: Archon Books, 1990.

Loras, Mathias. *Foundations: The Letters of Mathias Loras, D.D., Bishop of Dubuque.* Edited by Robert F. Klein. Dubuque, IA: Loras College Press, 2004.

Lorch, Fred W. "Iowa and the California Gold Rush." *Iowa Journal of History and Politics* 30, no. 3 (July 1932): 307–76.

Lyon, Randolph. *Dubuque: The Encyclopedia.* Dubuque, IA: [First National Bank?], 1991. Updated version continued as *Encyclopedia Dubuque* at www.encyclopediadubuque.org.

Macbride, Thomas H. "Forestry Notes for Dubuque County." Iowa Geological Survey. *Annual Report* 10 (1899): 623–51. Des Moines: Iowa Geological Survey, 1900.

Mahoney, Timothy R. *Provincial Lives: Middle-Class Experience in the Antebellum Middle West.* Cambridge, UK: University Press, 1999.

———. *River Towns in the Great West: The Structure of Provincial Urbanization in the American Midwest, 1820–1870.* Cambridge, UK: University Press, 1990.

———. "Urban History in a Regional Context: River Towns on the Upper Mississippi, 1840–1860." *Journal of American History* 72, no. 2 (September 1985): 318–39.

Manly, William Lewis. *Death Valley in '49. Important Chapter of California Pioneer History.* San Jose, CA: Pacific Tree and Vine Co., 1894. Reprint, Ann Arbor, MI: University Microfilms, 1966.

Map of the Mineral Region of Dubuque & Vicinity: State of Iowa. St. Louis, MO: L. Gast and Brother, 1858. Reprint, Dubuque, IA: Center for Dubuque History, Loras College, 1979.

Map of Oregon and Upper California from the Surveys of John Charles Fremont and other Authorities. Baltimore, MD: E. Weber, 1848. Published separately as a large folded map measuring 84 x 68 cm.

Maritime Heritage Project. maritimeheritage.org.

Mattes, Merrill J. "The Council Bluffs Road; Northern Branch of the Great Platte River Road." *Overland Journal* 3, no. 4 (Fall 1985): 30–42.

———. *Platte River Road Narratives: A Descriptive Bibliography of Travel Over the Great Central Overland Route to Oregon, California, Utah, Colorado, Montana, and Other Western States and Territories, 1812–1866.* Urbana: University of Illinois Press, 1988.

McKinstry, Byron Nathan. *The California Gold Rush Overland Diary, 1850–1852 with a Biographical Sketch and Comment on a Modern Tracing of His Overland Travel.* (American Trails Series, 10.) Glendale, CA: A.H. Clark Co., 1975.

Miners' Express [Dubuque, IA.]

Mintz, Lannon. *The Trail: A Bibliography of the Travelers on the Overland Trail to California, Oregon, Salt Lake City, and Montana during the Years 1841–1864.* Albuquerque: University of New Mexico Press, 1987.

Morton, J. Sterling, et al. *Illustrated History of Nebraska: A history of Nebraska from the Earliest Explorations of the Trans-Mississippi Region.* Lincoln, NE: J. North, 1905–13.

Murphy, Lucy Eldersveld. *Native American Lead Mining in the Upper Mississippi Valley.* In *Plumbing the Depths of the Upper Mississippi Valley: Julien Dubuque, Native Americans, and Lead Mining*. B. Pierre Lebeau, Lucy Eldersveld Murphy and Robert C. Wiederaenders, 5–19. (Extended Publications Series, 7.) Naperville, IL: Center for French Colonial Studies, 2008.

Newhall, John B. *A Glimpse of Iowa in 1846.* Iowa City: State Historical Society of Iowa, 1957.

———. *A New Map of Iowa for 1848–49, with Descriptive Notes: Designed for the Use of Emigrants and Travelers and as a Chart of Reference Invaluable to Every Citizen of the State.* Keosauqua, IA: Valley Whig Print, 1848.

Oldt, Franklin T. *History of Dubuque County, Iowa.* Chicago: Goodspeed Historical Association, 1911.

Otting, Loras C., ed. *Letters to a Pioneer Bishop; Correspondence to Mathias Loras, D.D., First Bishop of Dubuque.* Dubuque, IA: Loras College Press, 2009.

Owen, David Dale. *Report of a Geological Exploration of Part of Iowa, Wisconsin, and Illinois.* Washington, D.C.: Government Printing Office, 1844. 28th Congress, 1st session. Senate Document 407. Serial 437.

———. *Report of a Geological Survey of Wisconsin, Iowa, and Minnesota, and Incidentally of a Portion of Nebraska Territory, Made Under Instruction of the United States Treasury Department.* Philadelphia: Lippincott, Grambo & Co., 1852.

Parker, Nathan. *Iowa As It Is in 1855: A Gazetteer for Citizens, and a Hand-Book for Immigrants.* Chicago: Keen and Lee, 1855.

Parkman, Francis. *The California and Oregon Trail; Being Sketches of Prairie and Rocky Mountain Life.* New York: Putnam, 1849.

Petersen, William J. "Captain Daniel Smith Harris." *Iowa Journal of History and Politics* 28, no. 4 (October 1930): 505–42.

———. *Steamboating on the Upper Mississippi, the Water Way to Iowa: Some River History.* Iowa City: State Historical Society of Iowa, 1937. Reprinted by the Society in 1968.

Polk, James Knox. *Message from the President of the United States to the Two Houses of Congress at the Commencement of the Second Session of the Thirtieth Congress, December 5, 1848.* Washington, D.C.: 1848. 30th Congress, 2nd session. Executive Document No. 1, 56–64. Serial 537.

Preston, Howard Hall. *History of Banking in Iowa.* Iowa City: State Historical Society of Iowa, 1922.

Radisson, Pierre. "Radisson and Groseilliers in Wisconsin. The Third Voyage of Radisson." Wisconsin State Historical Society, *Collections* 11 (1888): 64–96. Madison: Democrat Printing Co., 1888.

Rasmussen, Louis J. *California Wagon Train Lists.* Vol. 1: April 5, 1849 to October 20, 1852. (A Volume of the Ship, Rail and Wagon Train Series). Colma, CA: San Francisco Historic Records, 1994.

Rodolf, Theodore. "Pioneering in the Wisconsin Lead Region." Wisconsin Historical Society, *Collections* 15 (1900): 338–89. Madison: Democrat Printing Co., 1900

Roth, Mitchel. "Cholera, Community, and Public Health in Gold Rush Sacramento and San Francisco." *Pacific Historical Review* 66, no. 4 (November 1997): 527–51.

Scharnau, Ralph. "From Pioneer Days to the Dawn of Industrial Relations: The Emergence of the Working Class in Dubuque, 1833–1855." *Annals of Iowa* 70, no. 1 (Summer 2011): 201–24.

Schockel, B.H. "Settlement and Development of the Lead and Zinc Mining Region of the Driftless Area with Special Attention to Jo Daviess County." *Mississippi Valley Historical Review* 4, no. 2 (1917–18): 169–92.

Schroder, Alan. *A Bibliography of Iowa Newspapers, 1836–1976.* Iowa City: State Historical Society of Iowa, 1979.

Shoemaker, [first name not identified]. *Overland Trail Diary*. Original manuscript diary at Special Collections, Harold B. Lee Library, Brigham Young University, Provo, Utah. contentdm.lib.byu.edu/u?/Diaries,4258.

Stillson, Richard Thomas. *Spreading the Word: A History of Information in the California Gold Rush.* Lincoln: University of Nebraska Press, 2006.

Storm, Colton. A *Catalogue of the Everett D. Graff Collection of Western Americana.* Chicago: Published for the Newberry Library by the University of Chicago Press, 1968.

Taylor, Henry. *From Lead Mines to Gold Fields: Memories of an Incredibly Long Life.* Edited and with an introduction by Donald L. Parman. Lincoln: University of Nebraska Press, 2006.

Taylor, Scott S. "Golden Letters: E.O.C. Ord and the California Gold Rush, 1848." *Manuscripts* 72, no. 2 (Spring 2010): 113–23.

Thwaites, Reuben Gold. "Early Lead Mining on the Upper Mississippi." In *How George Rogers Clarke Won the Northwest and Other Essays in Western History.* Chapter VII. Chicago: McClurg, 1903.

Turnbull, L. "The Mineral Contents of the Lower Magnesian Limestone of the Upper Mississippi. Condensed from Dr. Owen's Geological Survey of Wisconsin, etc." *Journal of the Franklin Institute* 27 (1854): 182–86.

U.S. *Public Statutes at Large of the United States of America*. "Articles of Treaty of Peace, Friendship, and Cession," 7, 374–76. Boston: Little, Brown, 1846.

U.S. *Statutes at Large and Treaties of the United States of America.* "Treaty of Peace, Friendship, Limits, and Settlement with the Republic of Mexico." 9, 922–42. Boston: Little, Brown, 1851.

Unruh, John D. *The Plains Across: The Overland Emigrants and the Trans-Mississippi West, 1840–60.* Urbana: University of Illinois Press, 1993.

Van der Zee, Jacob. "The Roads and Highways of Territorial Iowa." *Iowa Journal of History and Politics* 3, no. 2 (April 1905): 175–225.

Wagner, Henry R., and Charles L. Camp. *The Plains & the Rockies: A Critical Bibliography of Exploration, Adventure, and Travel in the American West, 1800–1865.* 4th edition. Revised, enlarged and edited by Robert H. Becker. San Francisco: John Howell Books, 1982.

West, Irma. "Plagues of the Gold Rush." *Sierra Sacramento Valley Medicine* 52, no. 4 (July–August 2001): 10–11.

"Western Towns. Dubuque in 1849." *Western Journal* 3, no. 5 (February 1850): 326–27.

Wheat, Carl I. *Mapping the Trans-Mississippi West, 1540–1861.* 5 vols. in 6. San Francisco: Institute of Historical Cartography, 1957–63.

Whitney, Josiah Dwight. *Report of a Geological Survey of the Upper Mississippi Lead Region.* Albany, NY: Made by Authority of the Legislature of Wisconsin, 1862.

Wilkie, William E. *Dubuque on the Mississippi, 1788–1988.* Dubuque, IA: Loras College Press, 1987.

INDEX

H

I

J

K

L

About the Author

Robert Klein is a lifelong resident of Dubuque who was educated in the local schools and graduated with a BA degree from Loras College. After tours in the U.S. Army, he received the MSLS degree from the Catholic University of America in Washington, D.C.

He then became assistant librarian at Loras College from 1964 to 1969 and library director from 1969 to 2004. He served for ten years on the Board of Directors and is a past president of the Dubuque County Historical Society. He is one of the co-founders of the Center for Dubuque History at Loras College.

He is the editor of two previous publications: Chandler Childs's *Dubuque: Frontier River City* (1984) and *Foundations: The Letters of Mathias Loras, Bishop of Dubuque* (2004).